COLOR SCIENCE AND THE VISUAL ARTS

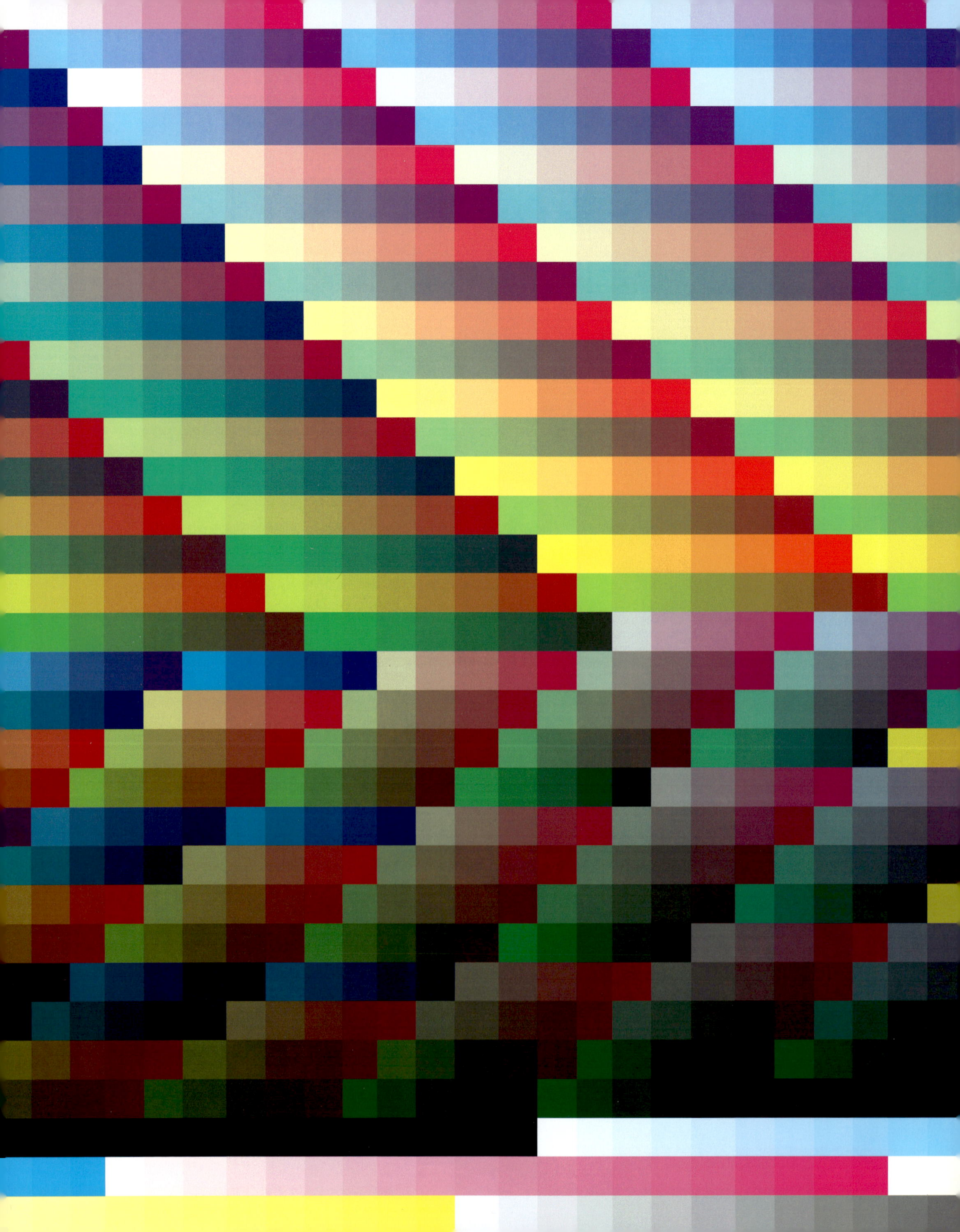

COLOR SCIENCE AND THE VISUAL ARTS

A GUIDE FOR CONSERVATORS, CURATORS, AND THE CURIOUS

ROY S. BERNS

The Getty Conservation Institute
Los Angeles

The Getty Conservation Institute
Timothy P. Whalen, *Director*
Jeanne Marie Teutonico, *Associate Director, Programs*

The Getty Conservation Institute works to advance conservation practice in the visual arts, broadly interpreted to include objects, collections, architecture, and sites. It serves the conservation community through scientific research, education and training, model field projects, and the broad dissemination of the results of both its own work and the work of others in the field. And in all its endeavors, it focuses on the creation and dissemination of knowledge that will benefit professionals and organizations responsible for the conservation of the world's cultural heritage.

Published by the Getty Conservation Institute
Getty Publications
1200 Getty Center Drive, Suite 500
Los Angeles, CA 90049-1682
www.getty.edu/publications

Beatrice Hohenegger, *Project Editor*
Katya Rice, *Manuscript Editor*
Jim Drobka, *Designer*
Stacy Miyagawa and Suzanne Watson, *Production*

Distributed in the United States and Canada by the University of Chicago Press

Distributed outside the United States and Canada by Yale University Press, London

Printed in China

Library of Congress Cataloging-in-Publication Data
Names: Berns, Roy S., 1954- author. | Getty Conservation Institute, issuing body.
Title: Color science and the visual arts : a guide for conservators, curators, and the curious / Roy S. Berns.
Description: Los Angeles, CA : Getty Conservation Institute, [2016] | © 2016 | Includes index.
Identifiers: LCCN 2015036618 | ISBN 9781606064818 (hardcover)
Subjects: LCSH: Art—Conservation and restoration. | Colorimetry. | Color in art. | Color vision. | Pigments.
Classification: LCC N8560 .B47 2016 | DDC 701/.85—dc23
LC record available at http://lccn.loc.gov/2015036618

Illustration Credits

Every effort has been made to contact the owners and photographers of objects reproduced here whose names do not appear in the captions or in the illustration credits listed below. Anyone having further information concerning copyright holders is asked to contact Getty Publications so this information can be included in future printings.

Unless otherwise noted, all illustrations are by Roy S. Berns.

Page 16: Susan L. Stanger
Figs. 1.7, 5.13: Van Gogh Museum, Amsterdam
Figs. 1.8, 4.8, 5.14, 6.16–6.17: Art Institute of Chicago
Figs. 2.4, 2.8–2.10, 2.19, 2.21–2.22, 2.24, 3.1, 3.14, 3.18, 5.4, 5.6, 6.11, 6.15, 7.1–7.3, 7.5, 7.7: J. Paul Getty Museum
Figs. 2.5, 6.5: Courtesy National Gallery of Art, Washington
Fig. 2.20a: Erich Lessing / Art Resource, NY
Fig. 2.20b: Erich Lessing / Art Resource, NY / Roy S. Berns
Fig. 4.5a: © 2015 Estate of Pablo Picasso / Artists Rights Society (ARS), New York
Fig. 5.13: Roy. S. Berns (Rochester Institute of Technology); Ella Hendriks, Teio Meedendorp, Louis van Tilborgh, and Maurice Tromp (Van Gogh Museum); Muriel Geldof and Maarten van Bommel (Netherlands Institute for Cultural Heritage)
Fig 5.14: Roy S. Berns and Brittany D. Cox (Rochester Institute of Technology); Kelly Keegan and Inge Fiedler (Art Institute of Chicago); Federica Pozzi (Guggenheim Museum); Maarten van Bommel (Netherlands Institute for Cultural Heritage)
Fig. 5.15: Brittany D. Cox and Roy S. Berns
Fig. 7.8: ICC Profile Swatch generated in Media Manager by Onyx Graphics Inc.

Title page: CMYK printer target (fig. 7.8, detail)

To the memories of Ethel Berns, Mark Gottsegen, Ross Merrill, and Max Saltzman, and to my wife, Susan Stanger, for the memories we have made and will make.

CONTENTS

FOREWORD

Color is everywhere. Understanding how we perceive it, and how that perception changes with different lighting conditions, as well as with age, through fading or other forms of discoloration, has clear consequences for how we view and treat works of art.

As part of its mission to advance conservation practice, the Getty Conservation Institute has often turned its attention to measuring color and developing ways of monitoring changes on works of art and methods to slow down those changes. These activities have included exploring improvements to measuring light sensitivity on works of art with micro-fadeometer instruments, investigating ways of measuring the transmitted light from fluorescent tubes, and assessing the benefits of anoxic environments on color change in paints. When Roy S. Berns shared his draft manuscript with us, we were struck by his ability to convey complex color technology concepts and solve problems that occur when art is displayed, conserved, imaged, or reproduced. So with the enthusiastic support of GCI scientists, the decision was made to publish *Color Science for the Visual Arts: A Guide for Conservators, Curators, and the Curious*.

Roy is a color scientist with degrees in textile science and chemistry. He wrote what many consider to be the definitive textbook on color technology, *Billmeyer and Saltzman's Principles of Color Technology, 3rd ed*. In his research, Roy's desire to understand how color science can improve the practice of creating, analyzing, displaying, reproducing, and conserving visual arts continued to grow and, in a natural progression, brought him in direct contact with conservation science.

During the last fifteen years, Roy has become extremely active in developing new imaging tools for the analysis of paintings, drawings, and artist materials. In 1999 he spent a year as a Visiting Senior Fellow in Conservation Science at the National Gallery of Art, Washington, DC, where he introduced instrument-based color matching to paintings conservation, analyzed digital photography color accuracy, and studied the optical properties of picture varnishes using imaging techniques. He has since worked with many of the world's major museums using a variety of imaging techniques to map paintings' materials across a surface and to create digital reconstructions of visual arts that have undergone undesirable color changes.

Roy spent six months at the GCI as a guest scholar in 2010, during which time he studied how conservators, art historians, and curators analyze and critique visual art, and how they evaluate and use art reproductions. The study was also an opportunity for him to identify ways in which the imaging tools he had developed in his conservation science research could be used to analyze visual art in new and unique ways, and to determine what new color science tools could be created to analyze visual arts and to simulate the display of art.

With his enthusiasm, keen mind, and depth of knowledge of color science, Roy was ideally positioned to provide a unique and valuable perspective and depth in the area of color science and art. We are pleased that his time at the GCI helped lay the groundwork for writing this book, and it is gratifying to see the final result in this helpful guide for "conservators, curators, and the curious."

Timothy P. Whalen
Director
The Getty Conservation Institute

PREFACE

A curator, a paintings conservator, a photographer, and a conservation scientist walk into a bar — their museum's coffee bar. The curator turns to the conservator and complains that the recent conservation treatment's imitative inpainting is a dreadful match. "When I was in the gallery late in the day, I was horrified," the curator says.

"I had the same problem," the photographer says, nodding. "The post-treatment image revealed the inpainting, although I couldn't see it on the painting myself. Fortunately, it was easy enough to retouch the colors for the official image."

"Something is not right," the curator adds. "As I was leaving the gallery, I turned to take a last look and couldn't see the mismatches. I went back for a closer look and there they were again."

The paintings conservator is surprised about both mismatches. "In the studio the inpainting was an excellent match. The gallery has natural lighting as well. The xenon lights in the photo studio are similar to daylight. I even had the painting analyzed to identify the pigments, and then I used the same ones."

All eyes turn now to the conservation scientist, who proceeds to describe—in great detail—the analytical procedures, how the pigments were identified, and the supporting literature, including author, journal, and year of publication.

A color scientist is at a nearby table eavesdropping. He walks over, introduces himself, and admits he was listening. He can explain the seeming conundrum, he tells the group. "Incandescent spotlights turn on to augment the natural daylight when the light levels get too low, right? The inpainting must be metameric: it will mismatch under the spotlights. If you had seen the painting without those," he says, turning toward the curator, "the inpainting would have been less visible."

Everyone at the table is listening intently. "Since the inpainting mismatches only in hue," the color scientist continues, "it quickly becomes invisible with distance because of our eye's spatial vision. I believe the correct pigments were identified, but their spectral properties have changed over time. Different pigments should have been used. Rather than matching the materials, the spectra should have been matched. The camera picked this up because it doesn't see color the same way we do. If you measure the inpainting and adjacent areas with a reflection spectrophotometer, you should see the spectra mismatching at longer visible wavelengths. It all comes down to metamerism."

There are some tentative nods. The color scientist tells the four that they might want to take a look at his new book, *Color Science and the Visual Arts: A Guide for Conservators, Curators, and the Curious*. "All of these topics are presented. There's even a chapter about metamerism."

Off they go to the museum gift shop, where four books leave the shelf, one for each department.

▪ ▪ ▪

My coffee-bar allegory encapsulates the purpose of this book: understanding and solving color-technology problems that occur when art is displayed, conserved, imaged, or reproduced. The book was written for two audiences. The first is museum professionals seeking explanations to common color-related problems. For example, chapter 7, "Color Reproduction," includes a section entitled "Why Reproducing Artwork Is Difficult" that could be used by photographers and production personnel to fully understand the color-matching problems between a painting and its reproduction in an exhibition catalogue. A conservation scientist could learn about defining colors and color differences numerically in chapter 3, "Visual and Numerical Color Descriptions."

The second audience is students in conservation, museum studies, and art history programs. Metamerism, a phenomenon in which two colors that match under one lighting mismatch under a different lighting, is one of many considerations when choosing colorants for imitative inpainting. Chapter 4, "Metamerism and Color Inconstancy," addresses this topic directly. Chapter 6, "Painting Materials," explains how light scattering both within and at the surface of a painting affects color. From chapter 5, "Displaying Artwork," a museum studies student would learn about lighting specifications and an art history student about how color changes over time. For conservators and art historians who rely on images of artwork, chapter 7, "Color Reproduction," explores issues related to image archiving and considers why images on the Internet and in books may not match the original artwork.

A unique feature of this book is the use of images as its main medium of communication. Color physics, color vision, and imaging science were used to produce visualizations throughout the book. The idea that color inconstancy—the unwanted change in color with changes in lighting—can depend on the artist's palette is visualized by comparing palettes containing either cobalt or Prussian blue in a Van Gogh painting. The effects caused by changes in the color and intensity of the light source are visualized for paintings by Monet and Toulouse-Lautrec. The aging of a natural varnish is shown for a Renoir painting. The complexity of vision is demonstrated on the book's cover.

Two kinds of graphs are used in the book. Spectral graphs, where some measure of light is plotted as a function of wavelength, are indispensable in explaining color mixing, the color of light sources, and metamerism, among other topics. Color graphs are plots of colors defined numerically and provide a way to interpret the numerical data. For example, the change in color as a natural varnish ages, the color accuracy of a digital camera, and why most printing systems cannot reproduce a modern artist's vibrant palette are more easily explained using color graphs.

The book contains only one mathematical formula. There are no molecular formulas or diagrams, no ray-tracing drawings of light propagating through a paint layer. This was a very deliberate decision. In their place are the visualizations, spectral graphs, and color graphs.

The chapters fall naturally into two sections, fundamentals and applications. Chapters 1–4, the fundamentals, cover the topics of spectral measurements, color and spatial vision, visual and numerical color descriptions, and metamerism and color inconstancy. Chapters 5–7, the applications, discuss artwork display, painting materials, and color reproduction.

Finally, "Further Reading" is an annotated bibliography of recommended books. Most are intended for in-depth study of a particular topic, such as the conservation of easel paintings, general and museum lighting, color reproduction, vision, and color science. Some of these books cover topics that are addressed in this volume as well, but they are written from different perspectives—those of chemistry, physics, or fine art and design. The other recommended books are edited volumes, with each chapter contributed by an expert in a specific field. The depth and perspective are chapter specific.

It would not be surprising if at this point you wondered, Why do we need another book about color and art? The answer is this book's unique perspective and depth. I am first and foremost a color scientist, a professor in one of the few graduate programs in color science in the world. My research spans color measurement, color tolerance formulas, metamerism and color inconstancy, optical models of paints and varnishes, spectral modeling of imaging devices, spectral and high-color-accuracy image capture and printing, 3-D rendering of paintings, and image archiving. I bring to the writing of this book my interactions with the museum community—conservators, conservation scientists, curators, and photographers—on a range of topics: lighting, paintings conservation, image archiving, and simulations of "un-aging" paintings and drawings.

This book is the result of one color scientist's world tour of the visual arts.

Roy S. Berns
Rochester, New York
July 2015

ACKNOWLEDGMENTS

I am extremely grateful to GCI for awarding me a Visiting Scholar Grant during 2010, providing me with the opportunity to focus on this book while surrounded by conservators, curators, scientists, librarians, photographers, and of course, artwork displayed in a unique setting. I am also grateful to GCI for publishing this book. Should all Visiting Scholars be this fortunate. GCI publications require a champion. I had two: Tom Learner, head of science, and Cynthia Godlewski, publications manager. Your support is greatly appreciated. During the editorial phase, Tom and Cynthia brought technical editor Janet Bridgland on board to help improve clarity. I am indebted to Janet: she guided me to write a book that far exceeded my goals. Katarina Rice, as copy editor, took the manuscript even further. Thank you, Katya. Project editor Beatrice Hohenegger guided the book through various stages. Designer Jim Drobka created a magnificent layout while production manager Karen Schmidt and production coordinators Stacy Miyagawa and Suzanne Watson insured excellent color management.

My family has supported me through this six-year process. My sister, Marla Berns, director of the Fowler Museum at UCLA, was my lifeline both for the history of art and the mysteries of museum life. My wife, Susan Stanger, was quick to quip, "Sounds great!" whenever I needed encouragement.

A lesson I learned at the beginning of my academic career is the importance of asking for help from colleagues and students:

- Scott Geffert, Paul Green-Armytage, Ella Hendriks, Joy Luke, Ashley Penna, Daniela Pinna, Joel Pokorny, Bruce Richards, Ming-ching James Shyu, Stanley Smith, E. Keats Webb, and Steven Weintraub reviewed one or more chapters.

- Farhad Abed, Lin Chen, Tongbo Chen, Brittany Cox, Maxim Derhak, Carl Salvaggio, Lawrence Taplin, and Dave Wyble helped with MATLAB programming.

- Mark Fairchild, Joel Pokorny, and Steven Shevell advised me on color vision and appearance.

- Rolf Kuehni shared his vast knowledge of color order systems and color history.

- Dave Wyble assisted with fluorescence terminology and data.

- Maxim Derhak, Ken Fleisher, Chris Gallagher, Scott Geffert, Phil Green, Garrett Johnson, Hideyasu Kuniba, Alan Newman, and Stanley Smith were outstanding sources of information on artwork imaging and color management.

- Robert Gamblin and Mark Gottsegen gave advice about painting techniques.

- Conservation scientists Jim Druzik, Tom Learner, Alan Phenix, and Greg Smith were generous with their knowledge and experience.

- Conservators Jim Coddington, Ella Hendriks, Yvonne Szafran, and Steven Weintraub were always helpful.

The RIT Munsell Color Science Laboratory continues to be a well I draw upon, relying on its equipment, my colleagues, and, most importantly, the inquisitive students with a passion for color science.

Roy S. Berns

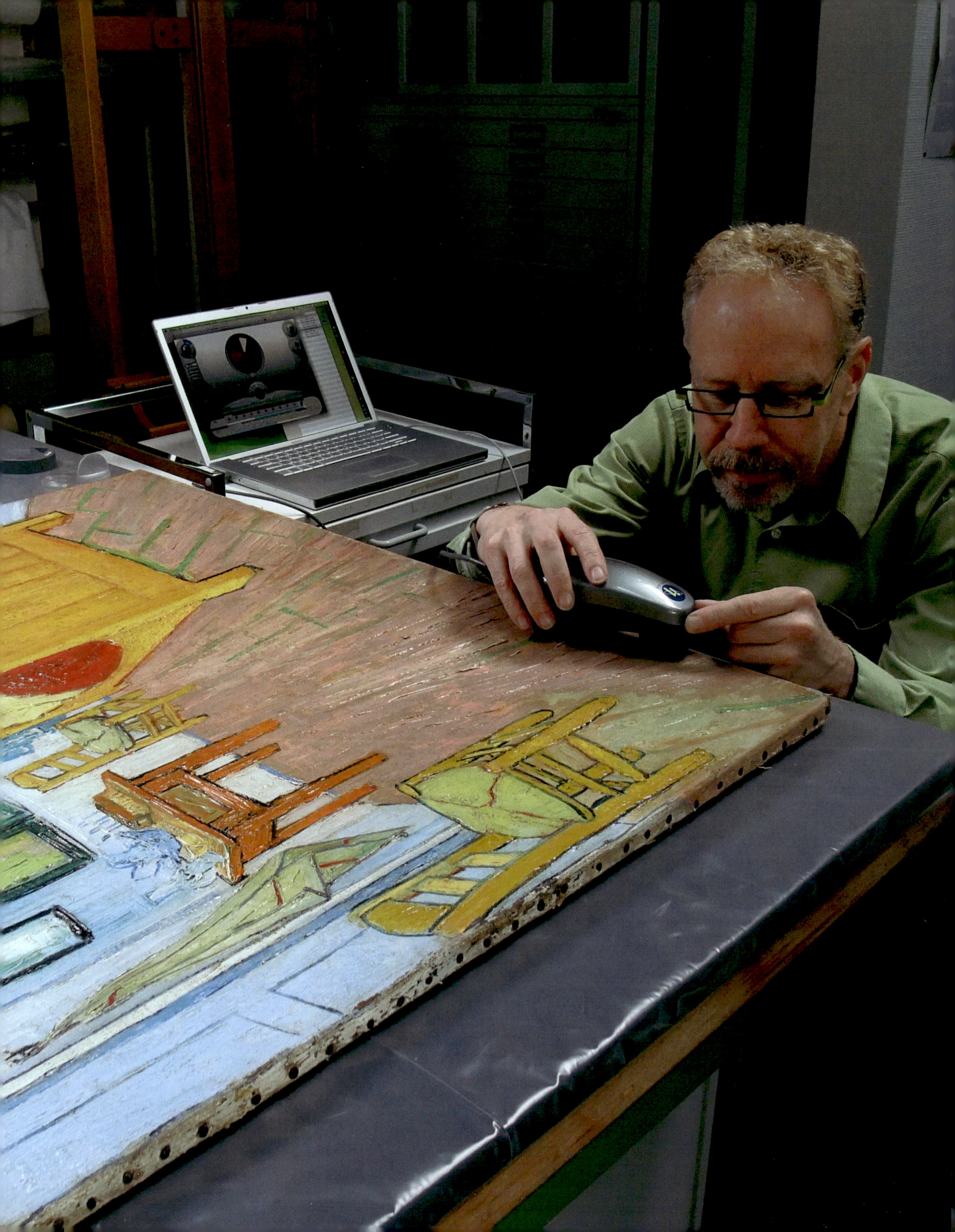

SPECTRAL MEASUREMENTS 1

Author using a reflection spectrophotometer to measure specific locations on Vincent van Gogh's *Bedroom*. The measurements were used to select pigments for inpainting.

Spectral measurement is one of many analytical techniques used in art conservation. For paintings and drawings, this technique, which involves graphing measured reflectance properties as a function of wavelength, can be used as an aid in colorant identification and authentication. Spectral data are also used for documentation: measurements taken over time can be analyzed to determine color changes caused by light exposure and identify which component in a mixture is fugitive (Johnston and Feller 1963). Information of this kind is critical for conservators and curators when balancing the need to preserve artwork with the desirability of making it available for public viewing (Michalski 1992). Spectral measurement is useful, too, for evaluating new conservation techniques that can affect the artwork's color and for ensuring consistency in the manufacture of artists' paints. Light sources are also measured spectrally. Most commonly, lighting is measured. In some cases, the artwork itself is measured, such as a piece made up of colored fluorescent lamps. Finally, spectral measurements can be taken of a single point or, through imaging, of the entire object.

The Visible Spectrum

In the seventeenth century Newton and others (Hunt 2000) demonstrated that when a narrow beam of sunlight was passed through a prism in a darkened room, the white-appearing light was dispersed into a spectrum, as shown in figure 1.1. When one of the colored beams was passed through a second prism, the beam was not further dispersed; the identical color persisted. Thus, these individual beams are unique and can be used to define and distinguish colored materials and lights.

It is convenient to define these beams as waves of a specific length and intensity. Consider the beams from green and red laser pointers, plotted in figure 1.2. Each of the lasers emits a single wavelength of light. The green laser emits a shorter wavelength—that is, having a shorter distance between peaks—than the red laser. The green laser also has twice the intensity; its maximum height is twice that of the red laser.[1]

Our environment is filled with waves, produced naturally (such as by the sun) or electronically (such as by a cell-phone transmitter). The waves we see are called light and range from about 380 to 780 nanometers, abbreviated "nm," where one nanometer is 1/1,000,000,000 meter (one billionth of a meter). The specific range of "visible" wavelengths varies from person to person and from instrument to instrument. Ultraviolet (UV) radiation occurs at shorter wavelengths than the visible region, and infrared (IR) radiation occurs at longer wavelengths than the visible region.

Intuitively, we know that lights of different colors must have different intensities at each wavelength within the visible spectrum. When lighting is measured, intensity is defined using units of radiance and can be graphed with wavelength on the horizontal axis and radiance on the vertical axis.[2] The two laser beams of figure 1.2 are plotted using this kind of graph, shown as figure 1.3.

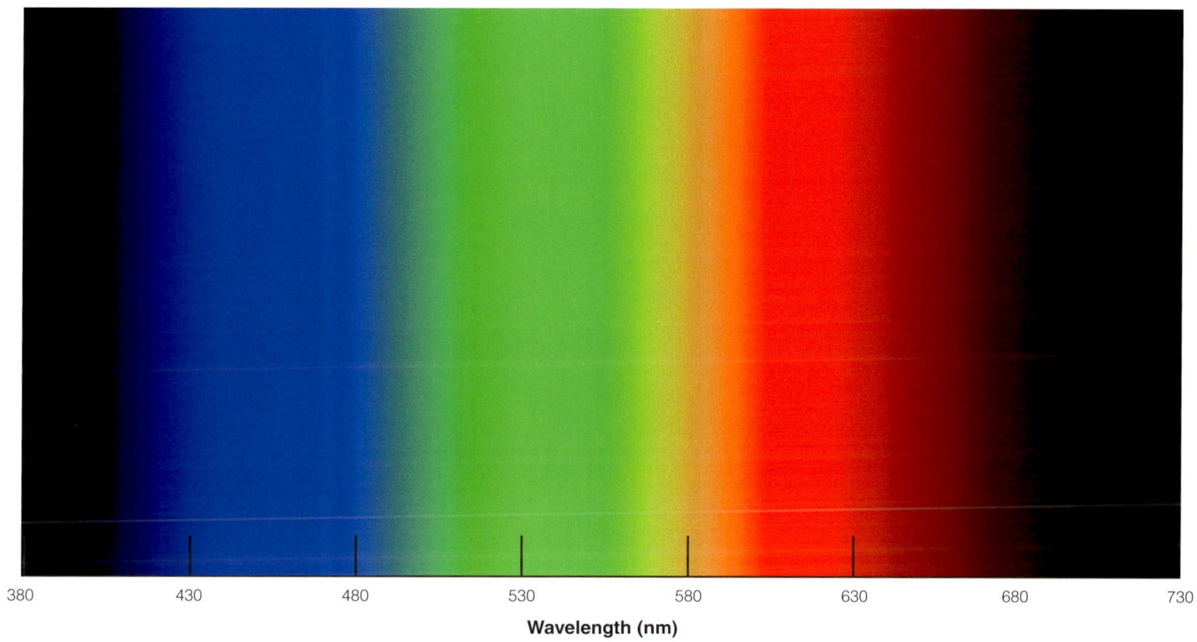

FIGURE 1.1. A visualization of dispersing sunlight into the spectrum.

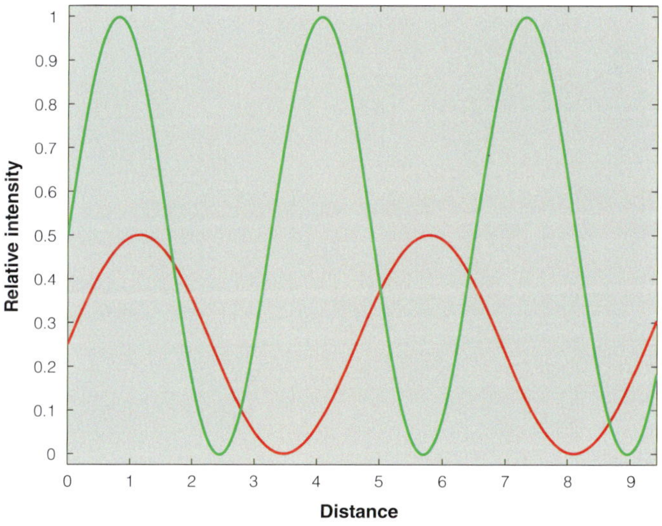

FIGURE 1.2. Light can be defined as a wave, where the distance from peak to peak is its wavelength. These are the spectra of two lasers with different wavelengths and intensities. The more intense laser has a higher relative intensity at its peak.

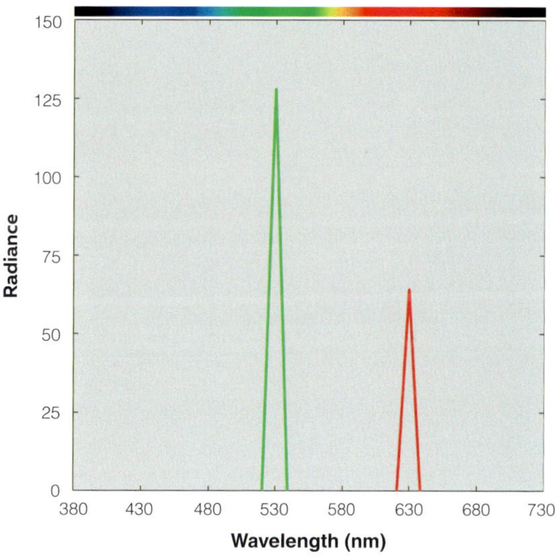

FIGURE 1.3. Spectral plot of radiance as a function of wavelength for the green and red lasers plotted in figure 1.2.

Measuring Light with a Spectroradiometer

Spectroradiometers are instruments that measure radiance as a function of wavelength. Such an instrument was used to measure the spectral properties of the two laser pointers. Spectroradiometers have components that focus the light, disperse it into individual wavelengths, and record each wavelength's intensity. The most versatile spectroradiometer uses a camera lens to collect and focus the light and has a circular aperture, which we see through the eyepiece as a black dot, to limit the light, thus forming a narrow beam of light.[3] This light strikes a diffraction grating, a piece of glass with hundreds of finely ruled lines, which disperses it into a spectrum. The spectrum lands onto an array of detectors where each detector corresponds to a wavelength. An analog-to-digital converter turns the detector-array electrical signals into digital signals. Following an internal calibration, data are output as a table or a graph of radiance versus wavelength (as in figure 1.3).

For figure 1.4, measurements were taken outdoors on a sunny day and indoors lit by a 60W incandescent light bulb. Because there is much

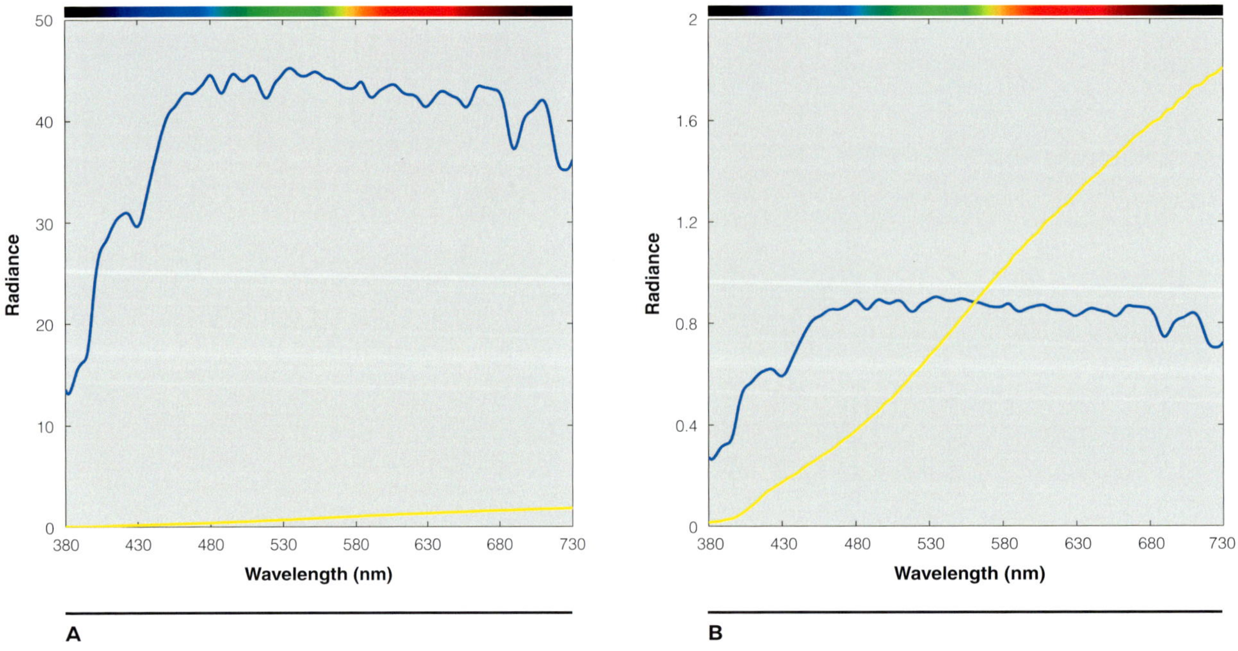

FIGURE 1.4. Spectra of outdoor sunlight (blue line) and a 60W incandescent light (yellow line), (A) before and (B) after scaling to appear equally bright.

more light outdoors than indoors, the light bulb's spectrum seems nearly flat. To reveal differences in their spectra, the two measurements were rescaled in such a way that the sources would appear equally bright if viewed side by side. Plotted in this manner, the measurements reveal the spectral signatures of the two kinds of light. As compared with incandescent light, daylight is composed of much greater amounts of UV radiation (shorter than 380 nm) and blue light, and at longer wavelengths incandescent has much greater radiance. These differences have a profound effect on the appearance of colors in artwork and the potential for light-induced damage (described in detail in chapter 5).

Measuring Materials with a Spectrophotometer

Materials are colored because they modify the light that strikes them, often a result of absorption and scattering, though there are many ways to produce color (Nassau 2001; Tilley 2011). Spectrophotometers are instruments that measure a material's reflectance or transmittance as a function of wavelength. Components include a light source, an optical system that focuses the light onto the material and collects the light that reflects from or transmits through the material, and—as in a spectroradiometer—a diffraction grating, detector array, and digital converter. Measurements of a white reflection standard, air, or a cuvette filled with solvent are used for calibration.

Reflectance and transmittance are measurements of the ratio of light reflecting from or transmitting through a material relative to the amount of incident light. If the light source is made dimmer, the ratio will not change.[4] Thus, we are measuring only the properties of the material, not the resulting interaction of the light source and the material. (The exception is fluorescent materials; the amount of fluorescence depends on the light source, as explained in chapter 6.)

Two tint ladders were produced using either ultramarine or cobalt blue and titanium dioxide white matte acrylic dispersion paints and measured with a spectrophotometer.[5] The results are shown in figures 1.5 and 1.6. Notice that "reflectance factor" is plotted rather than "percent reflectance." Spectrophotometers designed for color measurement are calibrated relative to a white material that reflects all incident light, and "factor" clarifies this calibration. For simplicity, however, "factor" has been omitted from the text. There are systematic changes in reflectance from

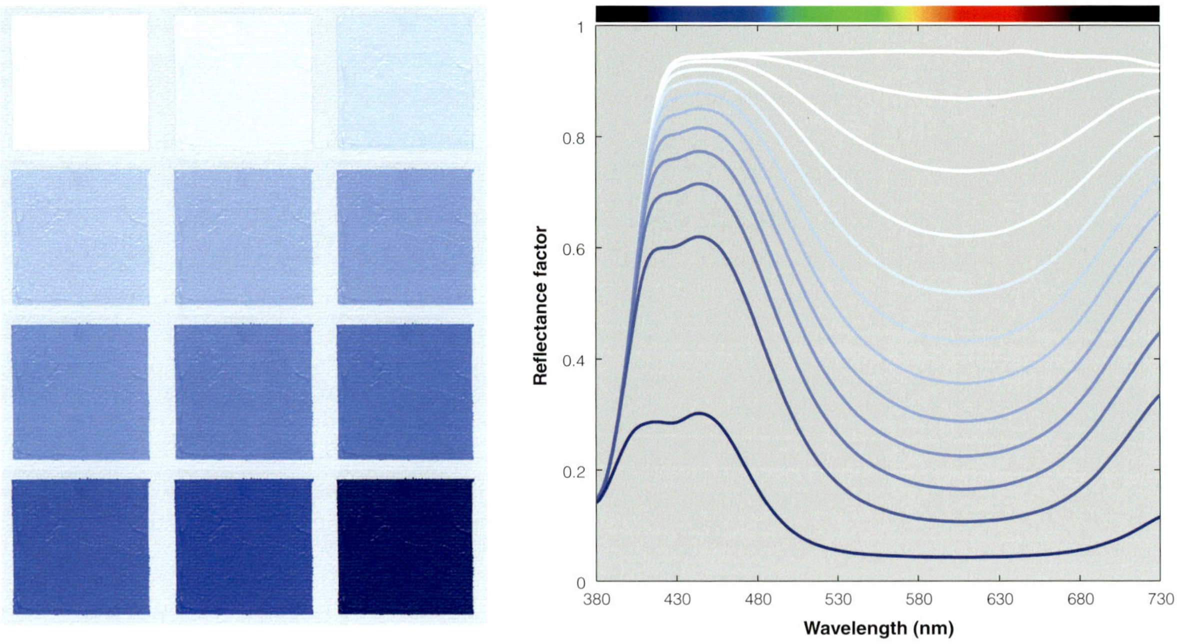

A **B**

FIGURE 1.5. (A) Tint ladder of titanium white and ultramarine blue paints, with (B) its reflectance spectra. Increasing the concentration of blue increases the amount of light absorbed, reducing reflectance.

100% white paint to 100% blue paint, where less light is reflected as the percentage of blue paint increases. The reduction in reflectance is a result of an increase in absorption. The two blues have very different spectra, particularly at long wavelengths.

The first spectrophotometers constructed specifically to measure paintings were very large and cumbersome (Wassall and Wright 1973; Wright 1981). Today, there are a variety of hand-held spectrophotometers well suited to measuring artwork, such as the one shown on p. 16. Vincent van Gogh painted three versions of his bedroom while in Arles, two of which were analyzed spectrally. The two—figures 1.7 and 1.8—are part of the permanent collections of the Van Gogh Museum in Amsterdam and the Art Institute of Chicago. The Van Gogh Museum's *Bedroom* contains three blues: cobalt, ultramarine, and Prussian blue, determined using X-ray diffraction, microraman spectroscopy, and X-ray fluorescence, respectively (Geldof, Megens, and Salvant 2013). It was of interest to see if a hand-held

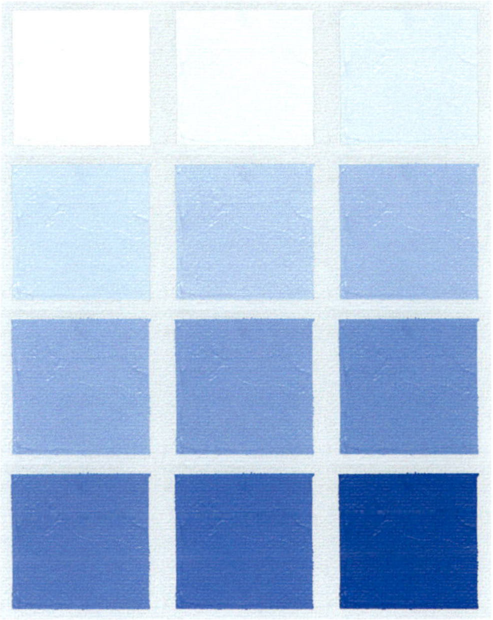

A

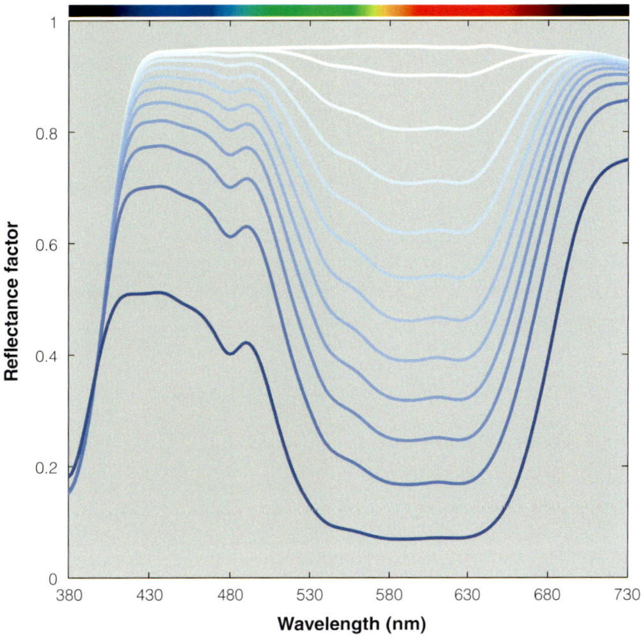

B

FIGURE 1.6. (A) Tint ladder of titanium white and cobalt blue paints, with (B) its reflectance spectra. The high reflectance at very long wavelengths and the mitt-like shape at short wavelengths are characteristic for cobalt blue, irrespective of binding medium.

spectrophotometer could be used for identification, particularly because it is a fraction of the cost of other analytical instruments. A graph of the spectral reflectance of the top left-hand door from both paintings is shown in figure 1.9. Both spectra have high reflectance at long wavelengths. The tint-ladder spectra of cobalt and ultramarine blue are reference spectra. Comparing the painting and reference spectra clearly indicates cobalt blue in both paintings. The Van Gogh Museum painting has a lower reflectance at short wavelengths, a result of short-wavelength absorption from a yellowed, seventy-year-old natural varnish. The varnish of the Art Institute of Chicago's version is colorless. The left-hand door in both paintings and the outline around the green window frame in the Art Institute's version are a deep blue. The spectral measurement of the window-frame outline identifies the blue as ultramarine. Van Gogh was very deliberate in his use of both cobalt and ultramarine blue.

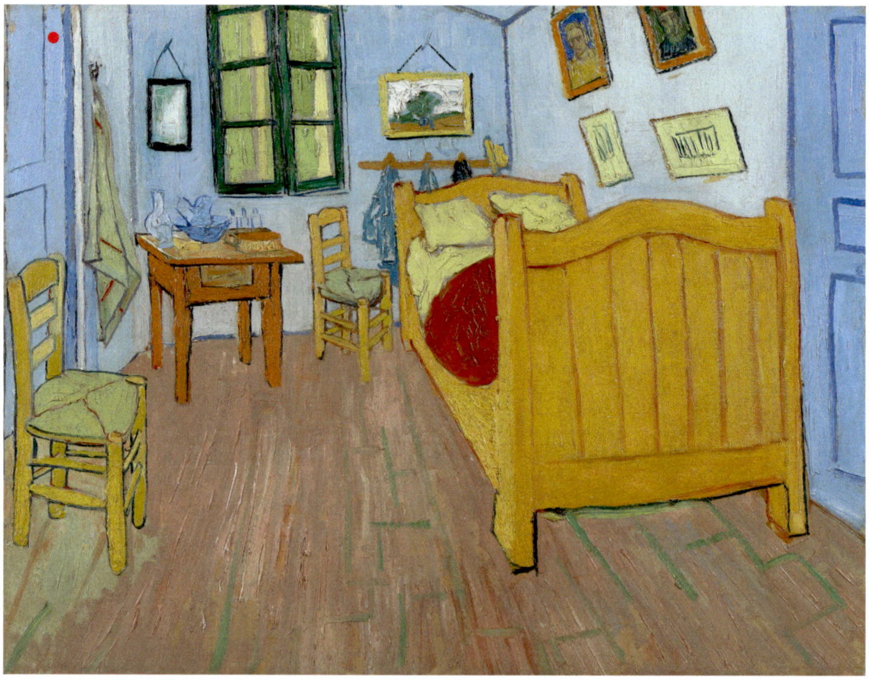

FIGURE 1.7. Vincent van Gogh (Dutch, 1853–1890), *Bedroom,* 1888. Oil on canvas, 72.4 × 91.4 cm (28½ × 36 in.). Amsterdam, Van Gogh Museum. Measurement position is shown as a red dot.

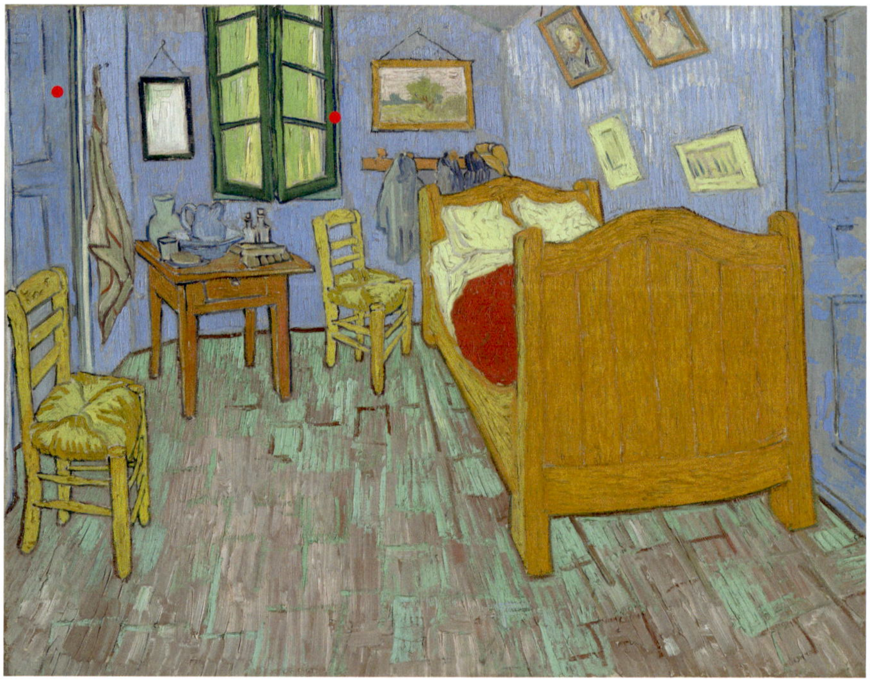

FIGURE 1.8. Vincent van Gogh (Dutch, 1853–1890), *Bedroom,* 1889. Oil on canvas, 73.7 × 93 cm (29 × 36⅝ in.). Art Institute of Chicago, 1926.417. Measurement positions are shown as red dots.

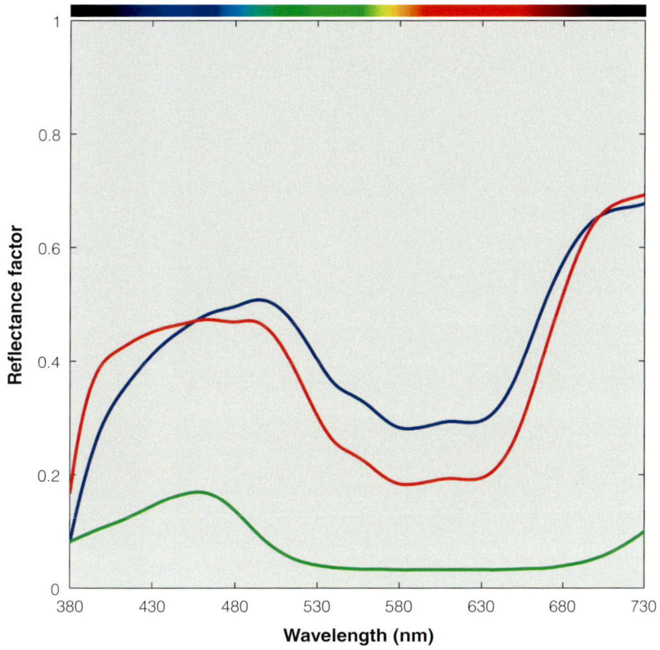

FIGURE 1.9. Spectral reflectance measurements of the left-hand door of the Van Gogh Museum's *Bedroom* (blue line) and the left-hand door (red line) and blue around frame (green line) of the Art Institute of Chicago's *Bedroom*.

Because of the effectiveness of this technique, libraries of reference spectra for various artist colorants appeared as early as 1939 (Barnes 1939), culminating in 2001 (Johnston-Feller 2001). One limitation of this approach is that the spectral signature varies with concentration, making identification difficult, particularly for very dark colorants. This is remedied by transforming the reflectance data using logarithms or derivatives, resulting in spectral signatures that are invariant to changes in concentration (Derby 1952; Johnston and Feller 1963; Bacci 1995; Johnston-Feller 2001).

Measuring Materials with Multi- and Hyperspectral Imaging

The basic concepts of spectrophotometry have been extended into the domain of digital imaging, where a camera replaces the detector array; this is often called multispectral or hyperspectral imaging, depending on the number of wavelengths that have been sampled. Such images can be used for technical examinations, color reproduction, and pigment

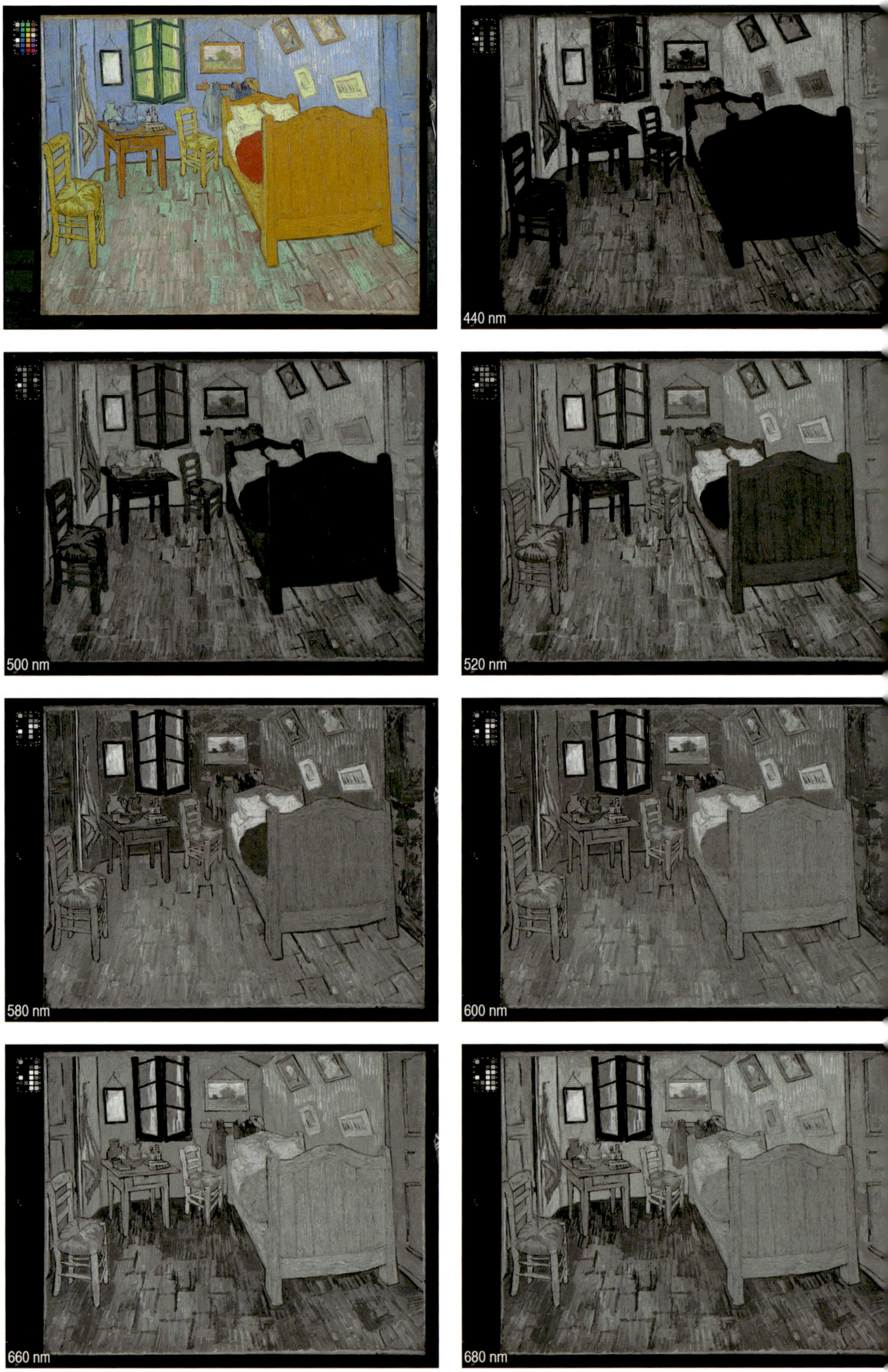

FIGURE 1.10. Multispectral images of the Art Institute of Chicago's *Bedroom* shown from 440 nm to 720 nm in 20 nm increments. The grayscale encoding corresponds to spectral reflectance ranging from zero (black) to unity (white).

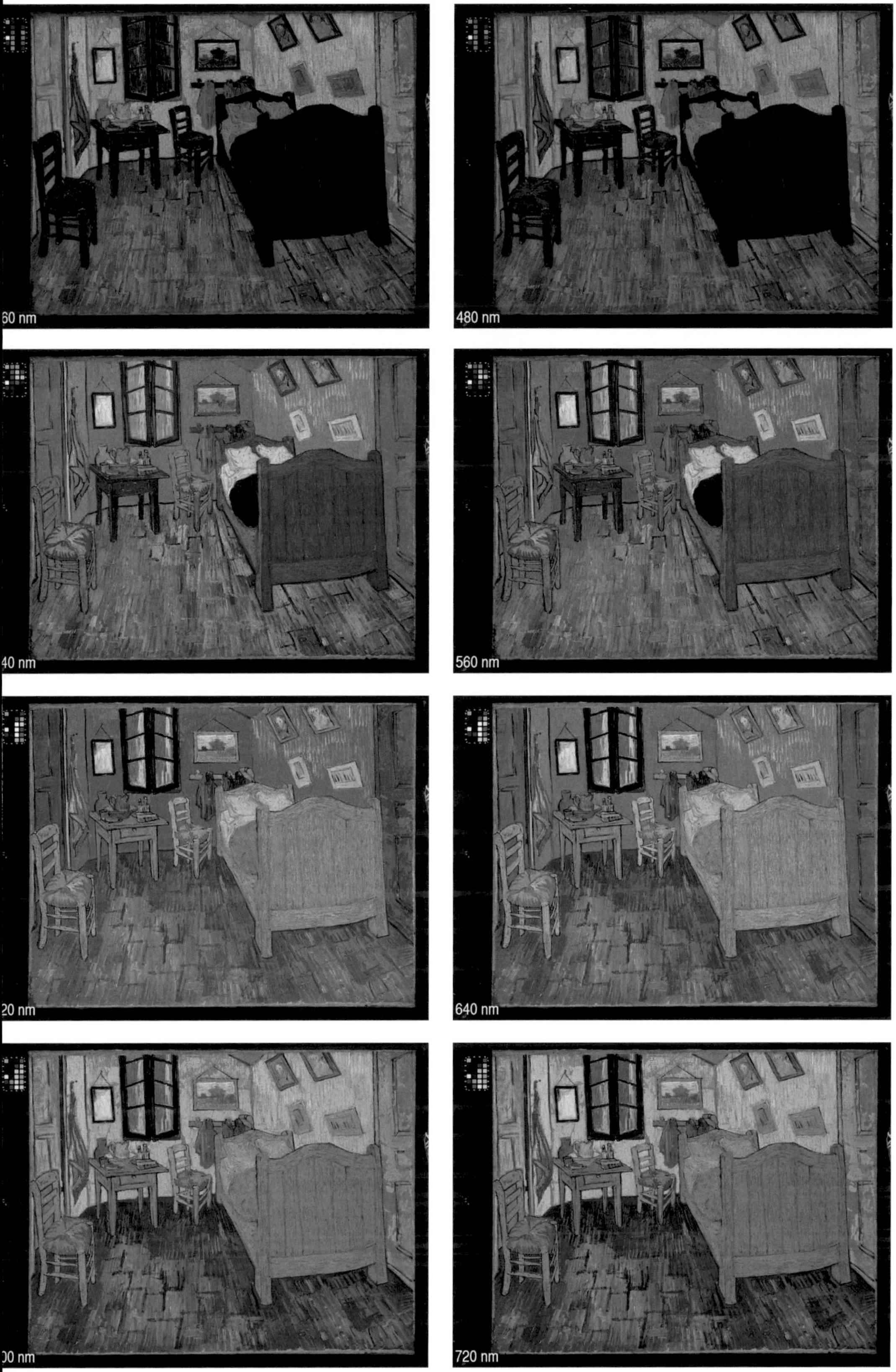
480 nm
560 nm
640 nm
720 nm

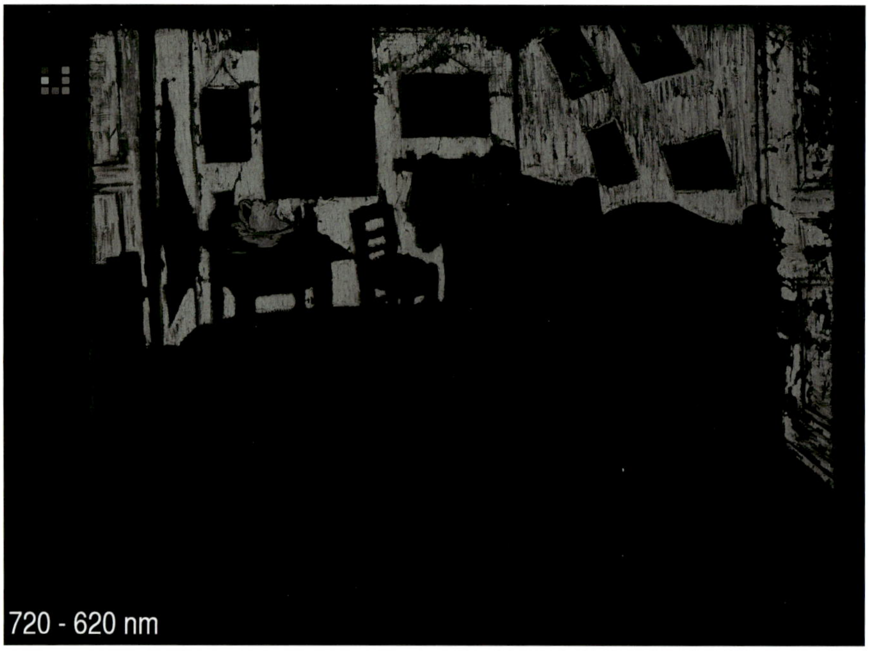

FIGURE 1.11. Subtracting the 620 nm image from the 720 nm image results in a mapping of cobalt blue for the Art Institute of Chicago's *Bedroom*.

mapping, where the spectral image data are transformed to concentration maps, providing new insights into an artist's working methods (Fischer and Kakoulli 2006; Pelagotti et al. 2008; Bianco et al. 2011). A multispectral system (Berns, Taplin, et al. 2005) was used to image the Art Institute of Chicago's *Bedroom,* resulting in thirty-six images corresponding to a range of 380 nm to 730 nm in 10 nm increments, a number of which are shown in figure 1.10 (on previous spread). Each wavelength image is grayscale, where black and white correspond to zero and unity reflectance, respectively. At short wavelengths, yellows and reds have low reflectance, while blues have high reflectance. With increasing wavelength, yellows, then oranges, then reds increase in reflectance. At the longest wavelengths, blues have high reflectance. As is clear in figure 1.6, cobalt blue absorbs light at 620 nm and reflects light at 720 nm. Subtracting the 620 nm image from the 720 nm image results in the mapping of cobalt blue that is shown in figure 1.11. This simple approach is effective because the walls and doors are predominantly cobalt blue. When a painting's colors are mixtures, more complex approaches are required for mapping (Abed 2014).

Summary

This chapter described the fundamentals of measuring the spectral properties of lights and materials using a spectroradiometer, spectrophotometer, or multispectral camera. An example was presented that used spectral data to identify cobalt and ultramarine blues in two versions of Van Gogh's *Bedroom*.

Spectral measurements are used to characterize the physical properties of materials and lights. Of equal importance to the physical measurements is how our visual system interprets these spectra, considered in chapters 2 and 3.

Notes

1. The green laser was assigned an arbitrary intensity of 1.0. The red laser has an intensity that is 50% of the green laser; thus its relative intensity is 0.5.
2. Radiance has units of milliwatts per steradian per meter squared per nanometer ($mWsr^{-1}m^{-2}nm^{-1}$). A milliwatt is 1/1,000 watt.
3. Such an instrument measures spectral radiance, a measurement that is distance independent. The other type of spectroradiometer measures spectral irradiance, a measurement that is distance dependent. Rather than a camera lens, there is a translucent diffuser, such as is used with light meters.
4. Because reflectance is a ratio, a spectroradiometer can be used to measure materials by illuminating the object with a light source with good temporal stability. First a white standard is measured, then the object, and finally their ratio is calculated. Such an approach has the advantage of not involving contact.
5. Samples "painted" for this book were produced computationally. An optical paint-mixing model was used to calculate spectral reflectance factor and CIELAB coordinates (Berns and Mohammadi 2007), and the coordinates were used to remap the color of an image of a painted square on a primed canvas support (Berns, Byrns, et al. 2006). This approach was more efficient and color accurate than producing actual painted squares and then taking digital photographs.

References

Abed, F. M. 2014. "Pigment Identification of Paintings Based on Kubelka-Munk Theory and Spectral Images." PhD diss. (Color Science), Rochester Institute of Technology.

Bacci, M. 1995. "Fibre Optics Applications to Works of Art." *Sensors and Actuators, B, Chemical* B29 (1–3): 190–96.

Barnes, N. F. 1939. "A Spectrophotometric Study of Artists' Pigments." *Technical Studies in the Field of the Fine Arts* 7: 120–38.

Berns, R. S., S. Byrns, F. Casadio, I. Fiedler, C. Gallagher, F. H. Imai, A. Newman, and L. A. Taplin. 2006. "Rejuvenating the Color Palette of Georges Seurat's 'A Sunday on La Grande Jatte – 1884': A Simulation." *Color Research and Application* 31: 278–93.

Berns, R. S., and M. Mohammadi. 2007. "Evaluating Single- and Two-Constant Kubelka-Munk Turbid Media Theory for Instrumental-Based Inpainting." *Studies in Conservation* 52: 299–314.

Berns, R. S., L. A. Taplin, M. Nezamabadi, M. Mohammadi, and Y. Zhao. 2005. "Spectral Imaging Using a Commercial Colour-Filter Array Digital Camera." In *14th Triennial Meeting, The Hague, 12–16 September 2005: Preprints (ICOM Committee for Conservation),* edited by I. Verger, 743–50. London: Earthscan.

Bianco, S., A. Colombo, F. Gasparini, R. Schettini, and S. Zuffi. 2011. "Applications of Spectral Imaging and Reproduction to Cultural Heritage." In *Digital Imaging for Cultural Heritage Preservation: Analysis, Restoration, and Reconstruction of Ancient Artworks,* edited by F. Stanco, S. Battiato, and G. Gallo, 183–214. Boca Raton, FL: CRC.

Derby, R. E. 1952. "Applied Spectrophotometry. I. Color Matching with the Aid of 'R' Cam." *American Dyestuffs Reporter* 41: 550–57.

Fischer, C., and I. Kakoulli. 2006. "Multispectral and Hyperspectral Imaging Technologies in Conservation: Current Research and Potential Applications." *Reviews in Conservation* 7: 3–16.

Geldof, M., L. Megens, and J. Salvant. 2013. "Van Gogh's Palette in Arles, Saint-Rémy and Auvers-sur-Oise." In *Van Gogh's Studio Practice,* edited by M. Vellekoop, M. Geldof, E. Hendriks, L. Jansen, and A. de Tagle, 238–55. New Haven, CT: Yale University Press.

Hunt, T. 2000. "Was Newton the First to Show That Visible Light Is Polychromatic?" *Engineering Science and Education Journal* 9: 185–91.

Johnston, R. M., and R. L. Feller. 1963. "The Use of Differential Spectral Curve Analysis in the Study of Museum Objects." *Dyestuffs* 44: 277–86.

Johnston-Feller, R. 2001. *Color Science in the Examination of Museum Objects: Nondestructive Procedures,* Tools for Conservation. Los Angeles: Getty Conservation Institute.

Michalski, S. 1992. "Sharing Responsibility for Conservation Decisions." In *Durability and Change: The Science, Responsibility, and Cost of Sustaining Cultural Heritage,* edited by W. E. Krumbein, P. Brimblecombe, D. E. Cosgrove, and S. Staniforth, 241–58. Chichester, UK: Wiley.

Nassau, K. 2001. *The Physics and Chemistry of Color: The Fifteen Causes of Color*. 2nd ed, Wiley Series in Pure and Applied Optics. New York: Wiley Interscience.

Pelagotti, A., A. D. Mastio, A. D. Rosa, and A. Piva. 2008. "Multispectral Imaging of Paintings: A Way to Material Identification." *IEEE Signal Processing Magazine* (July): 27–36.

Tilley, R. J. D. 2011. *Colour and the Optical Properties of Materials: An Exploration of the Relationship between Light, the Optical Properties of Materials and Colour*. 2nd ed. Chichester, UK: Wiley.

Wassall, M. P., and W. D. Wright. 1973. "A Special Purpose Spectrophotometer." In *Colour 73: Survey Lectures and Abstracts of the Papers Presented at the Second Congress of the International Colour Association,* edited by R. W. G. Hunt, 469–71. London: Adam Hilger.

Wright, W. D. 1981. "A Mobile Spectrophotometer for Art Conservation." *Color Research and Application* 6: 70–74.

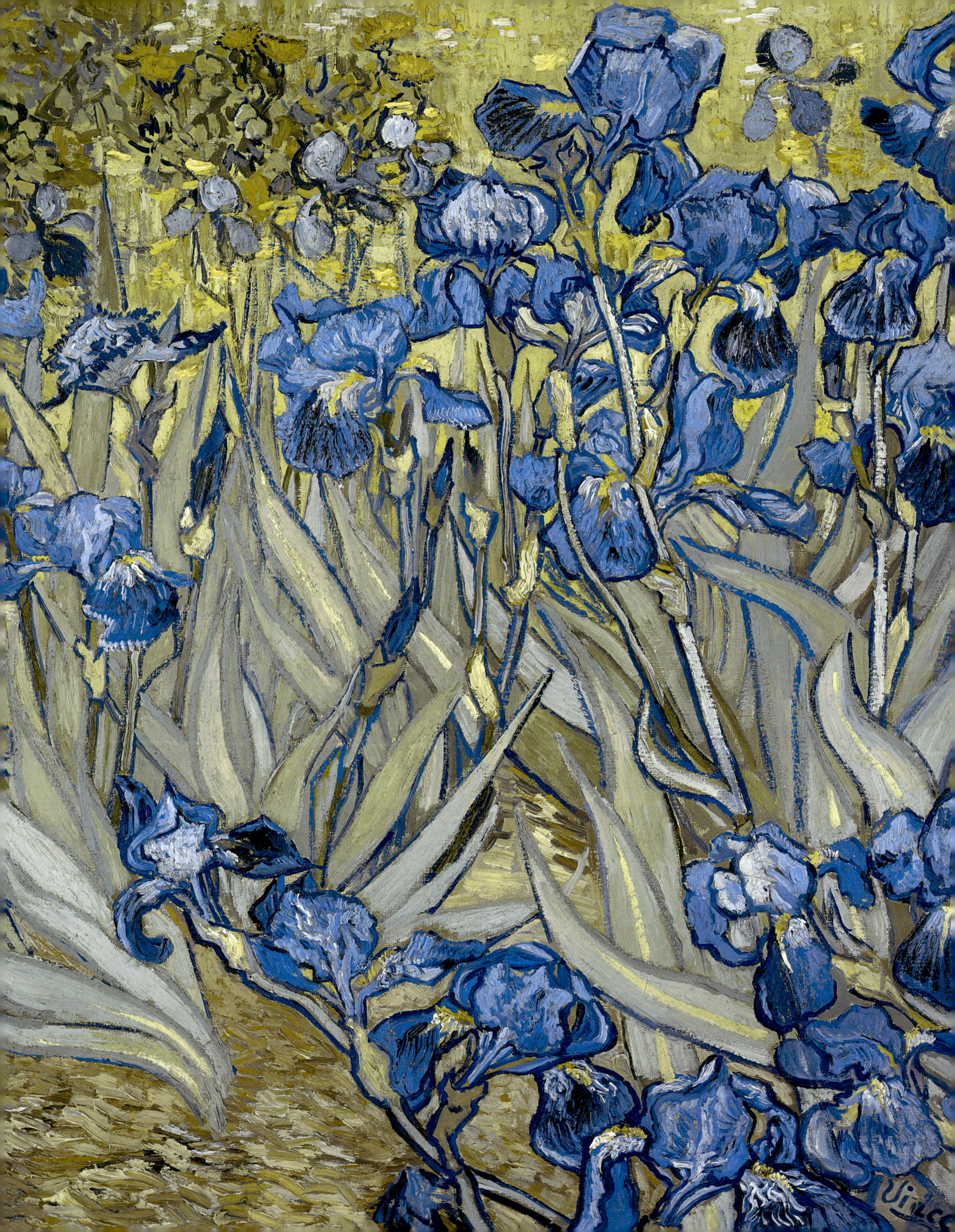

COLOR AND SPATIAL VISION

2

Converting a pattern of light that enters our eyes into a picture is exceedingly complex; it involves numerous physiological and psychological processes in our eyes and brain. Although different colorants in mixture lead to different spectral reflectances, they can result in the same color because of the physiology of our cone receptors. We may have a different interpretation of an artist's oeuvre if we know, for instance, that the artist developed cataracts over time. Using countershading to enhance depth and complementary colors to increase chromatic intensity results from the spatial interconnection of our receptors. JPEG file compression is often used to reduce file sizes of digital images. Visual artifacts are minimized successfully when the compression mimics both our color vision and our spatial vision, the subjects of this chapter.

Simulating the appearance of a detail of Vincent van Gogh's *Bedroom* for people with color-defective vision, in this case, so-called "red-green colorblind" (detail, fig. 2.10c)

Color Vision

How we see color has baffled and intrigued humankind for thousands of years. Early Greek philosophers believed that our eyes sent out rays rather than responding to light rays entering them (Lindberg 1976). Today we know that color occurs in the brain; it is the endpoint of a sequence of physical, physiological, and psychological events. But color is, in a sense, in the world as well: objects have color. Children are screened for color-vision deficiencies,[1] in part to prevent the misdiagnosis of a learning disorder because of the frequent use of color naming in early learning. Although it may seem paradoxical to state that color occurs in the brain, then, this is readily validated by our ability to produce the identical color sensation using many different combinations of colorants and coloring techniques. Color displays reproduce our chromatic world, composed of hundreds of colorants, with only three colored lights. Most printing systems use only four inks. If color were exclusively a physical property, materials would match only when their compositions were identical, thus producing identical spectral signatures.

Painters and color technologists know, however, that many different combinations of colorants can produce the same color. Neutral colors can be produced using black and white, complements such as blue and orange, or red, yellow, and blue. For figure 2.1, different acrylic dispersion paints were used to produce a matching gray pair when viewed under daylight.[2] One gray contained cobalt blue, arylide (hansa) yellow, quinacridone magenta, and titanium white. The other gray contained ultramarine blue, pyrrole orange, phthalocyanine green, and titanium white. Because very different paints were used to produce the two grays, their reflectance spectra are very different; this is particularly noticeable above 650 nm. Despite their dissimilar spectra, the same color sensation has been produced. This phenomenon, called metamerism, is considered in more detail in chapter 4.

A spectrophotometer has easily identified that the samples forming the pair were made using different colorants, but we ourselves, with our diminished spectral capacity, cannot distinguish these differences. The human visual system uses three sensors (called "cones," owing to their conical shape) for color vision. The spectrophotometer used for this book has thirty-six sensors that sample the visible spectrum from 380 nm to 730 nm in 10 nm increments. The most complex biological vision system is that of the mantis shrimp, a marine crustacean: it has between sixteen and twenty-one sensors, ten of which occur within our visible spectrum (Thoen et al. 2014). To survive and propagate the species, three color sensors and one night-vision sensor—which involves rods rather than cones—are sufficient for humans (Kevan and Backhaus 1998); for the mantis shrimp, however, many more are needed.

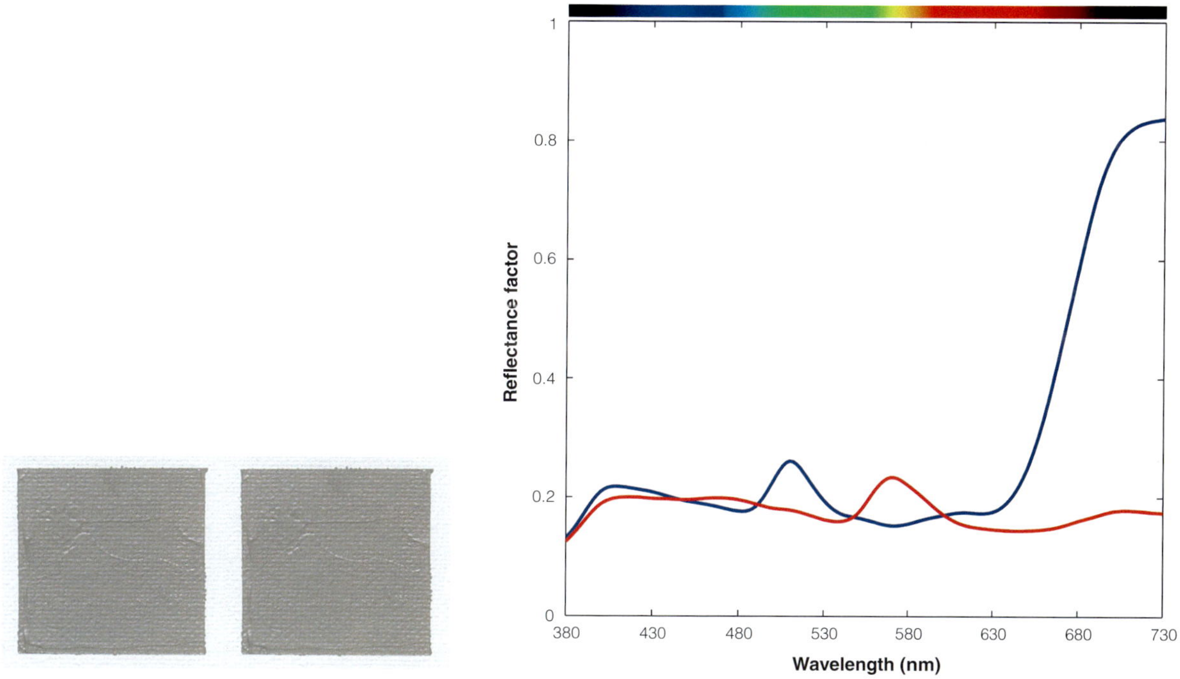

A

B

FIGURE 2.1. (A) A pair of color samples designed to match when illuminated with daylight, with (B) a graph of their reflectance spectra. The sample on the left (represented by the blue line in the graph) was made using cobalt blue (18% by weight), arylide yellow (63%), quinacridone magenta (10%), and titanium white (9%). The sample on the right (represented by the red line) contains ultramarine blue (36%), pyrrole orange (15%), phthalocyanine green (14%), and titanium white (35%).

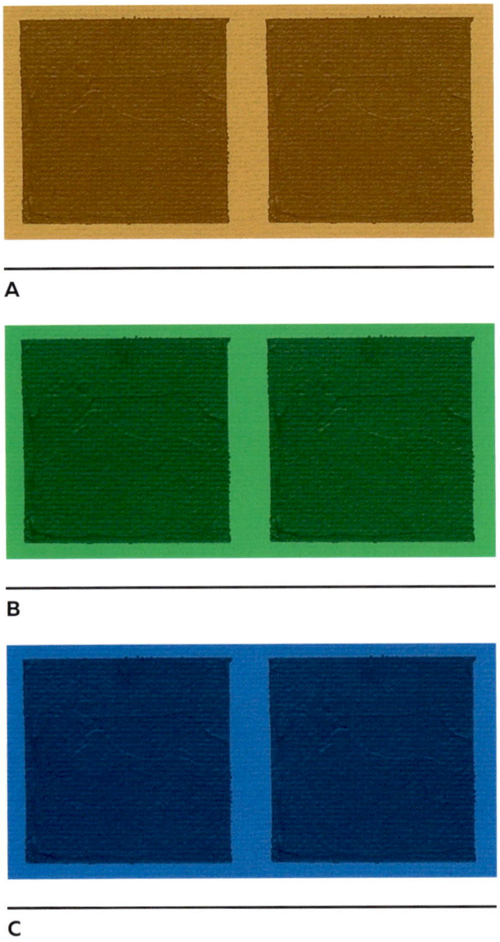

FIGURE 2.2. (A) Long (orange), (B) medium (green), and (C) short (blue) cone-receptor images of the matching gray pair when illuminated by daylight. These are "false color" images in which the specific hues are arbitrary. The key feature is the equal intensity of the two samples in each image.

Our visual system can be thought of as a camera that records three images, one from each of the cone types, as represented in figure 2.2. The two paints appear identical because the sensor signals are identical. Color vision is said to be trichromatic, and the three sensors are called the L, M, and S cone receptors, referring to long, medium, and short wavelength sensitivities. The spectra entering our visual system reduce to three signals; mathematically this is known as integration. Because of this integration, differences in reflectance cannot be discerned.

Creating figure 2.2 required knowing each color receptor's spectral sensitivity (that is, how the receptor responds to light, wavelength by wavelength), the spectral radiance of the lighting, and the spectral reflectance of each color patch. The spectral sensitivities of a typical color-normal observer are shown in figure 2.3. We do not sense the visible spectrum evenly; there is a gap between the S and M cones, and the L and M cones overlap considerably with sensitivities across much of the visible spectrum. It is inaccurate, then, to think of the L cones as sensing only red wavelengths, the M cones as sensing only green wavelengths, and the S cones as sensing only blue wavelengths.

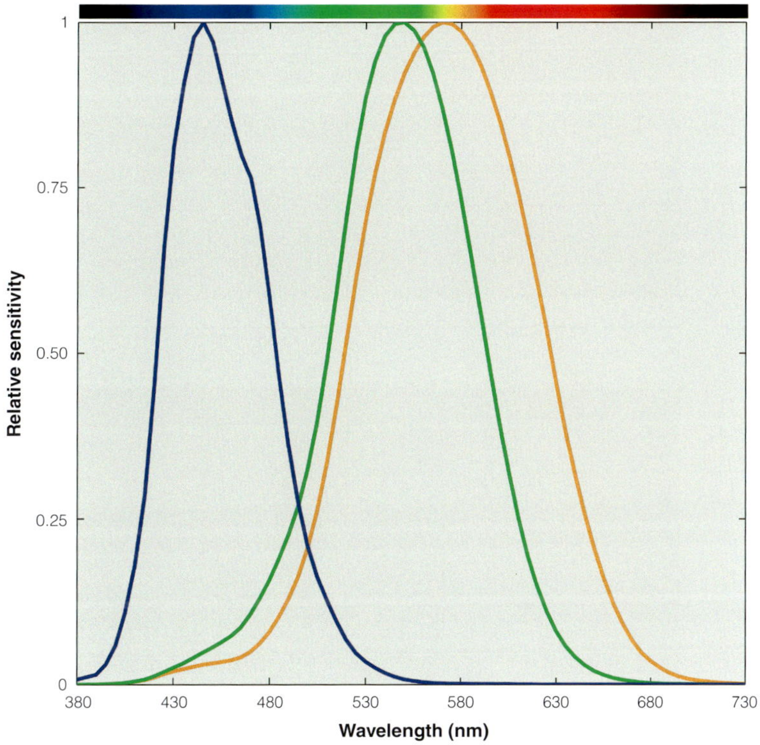

FIGURE 2.3. Spectral sensitivities of the human visual system cone receptors where the orange, green, and blue lines correspond to the L, M, and S cones (Smith and Pokorny 1975).

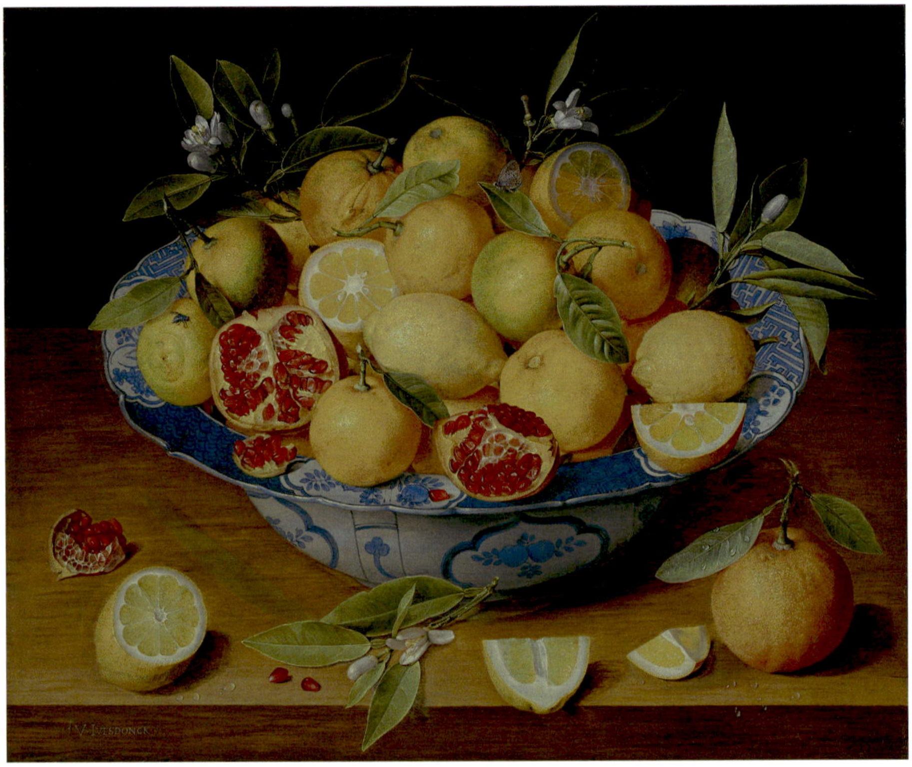

A

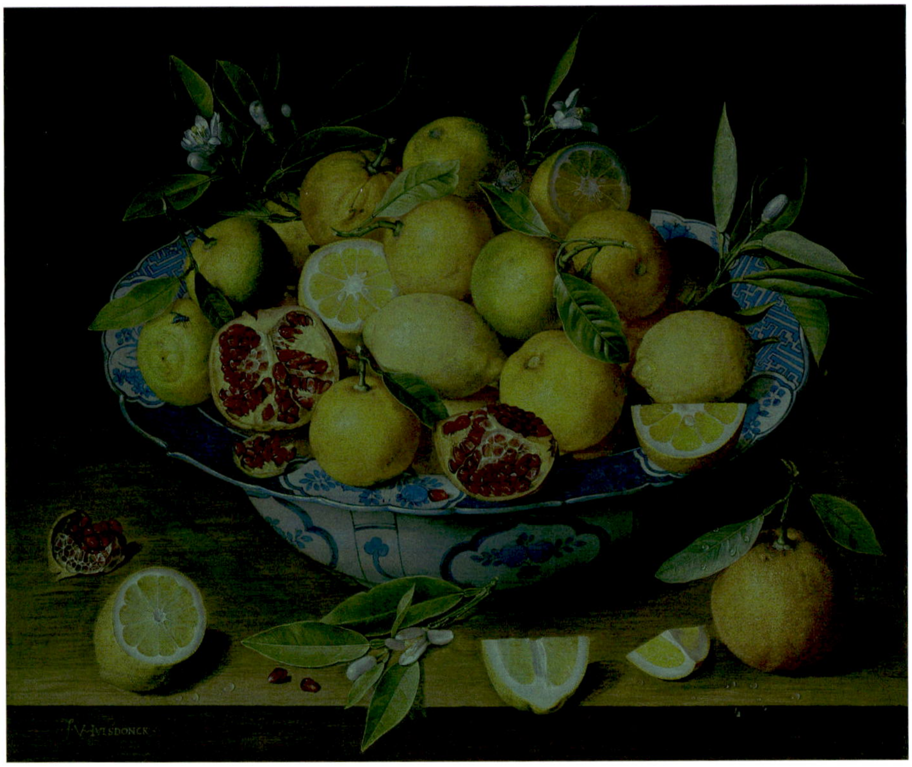

B

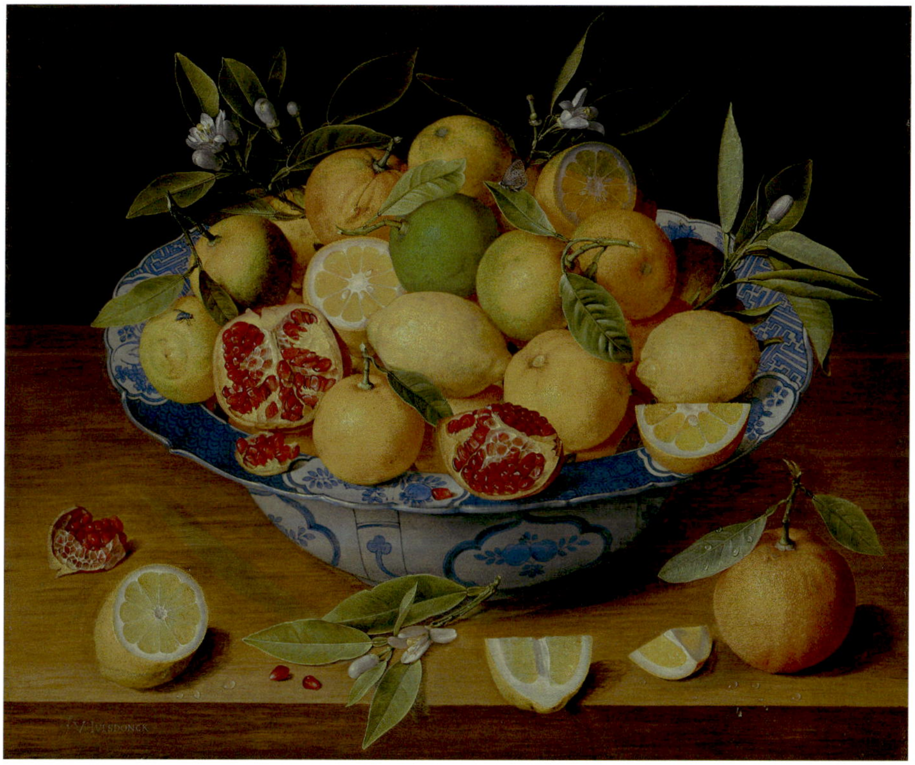

C

FIGURE 2.4. Jacob van Hulsdonck (Dutch, 1582–1647), *Still Life with Lemons, Oranges and a Pomegranate,* ca. 1620–1640. Oil on panel, 41.9 × 49.5 cm ($16\frac{1}{2} \times 19\frac{1}{2}$ in.). Los Angeles, J. Paul Getty Museum, 86.PB.538. The painting was imaged with (A) white and (B) colored light. When the central lemon from the colored-light image is overlaid on the white-light image (C), it becomes a lime, revealing our strong propensity to see fruit as color constant, even under colored light.

The trichromatic signals are not immediately sent along the visual pathway to our brain. They are first adjusted to compensate for the color of the lighting, a process known as chromatic adaptation. The adjustment is both physiological and psychological. Consider figure 2.4, in which *Still Life with Lemons, Oranges and a Pomegranate* by Jacob van Hulsdonck was imaged with white light and bluish light. Even in the colored lighting the fruit is still recognizable. Our visual system discounts the color of the lighting, aided by our expectation that colored objects, particularly food,

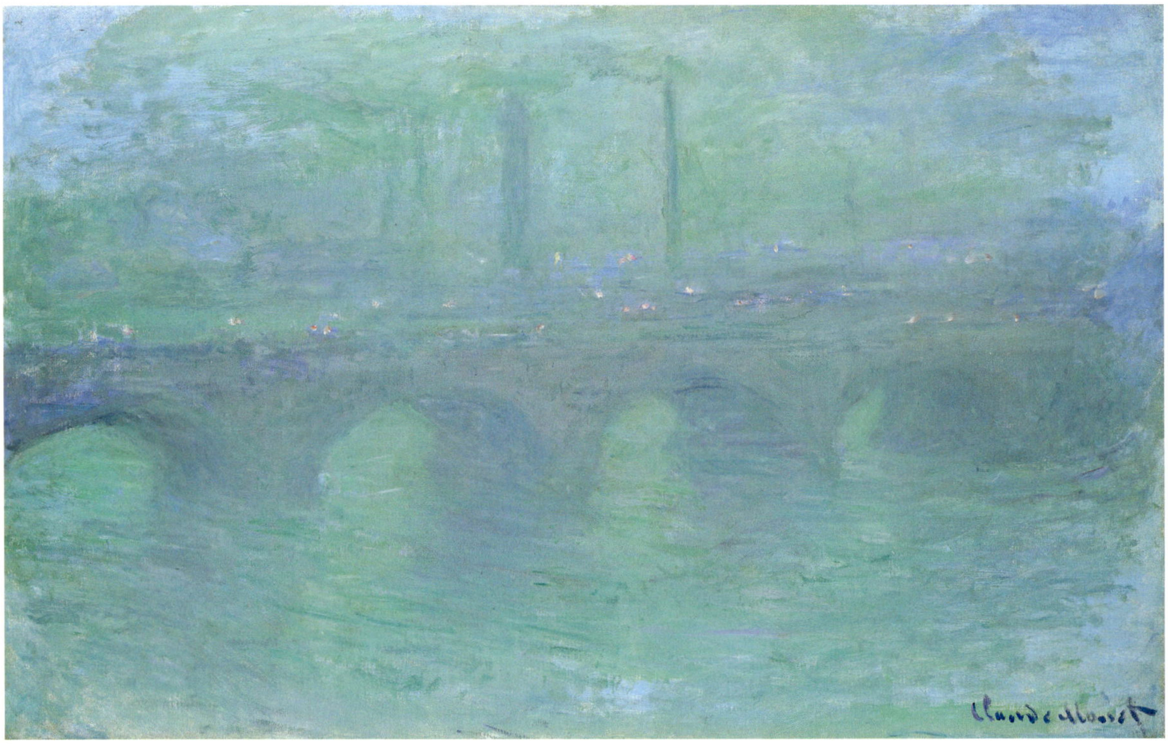

A

FIGURE 2.5. Claude Monet (French, 1840–1926). (A) *Waterloo Bridge, London, at Dusk*, 1904. Oil on canvas, 65.7 × 101.6 cm (25⅞ × 40 in.). Washington, National Gallery of Art, 19831.27. (B) *Waterloo Bridge, London, Sunset*, 1904. Oil on canvas 65.6 × 91.4 cm (25⅞ × 36½ in.). Washington, National Gallery of Art, 19831.28.

tend to appear constant in color. For the third image (fig. 2.4c), the central lemon from the blue-light image was overlaid on the white-light image. The lemon has become a lime!

Chromatic adaptation is sometimes incomplete, however. Candlelight and firelight produce orange light. A room lit by highly chromatic neon or solid-state lighting will never appear white. Painters are keenly aware that many types of lighting are chromatic, none more so than Claude Monet, who executed many paintings of a single subject under a wide range of conditions. The degree to which the color of natural light can vary with time of day, weather, and season is made clear in figure 2.5, two paintings of London's Waterloo Bridge.

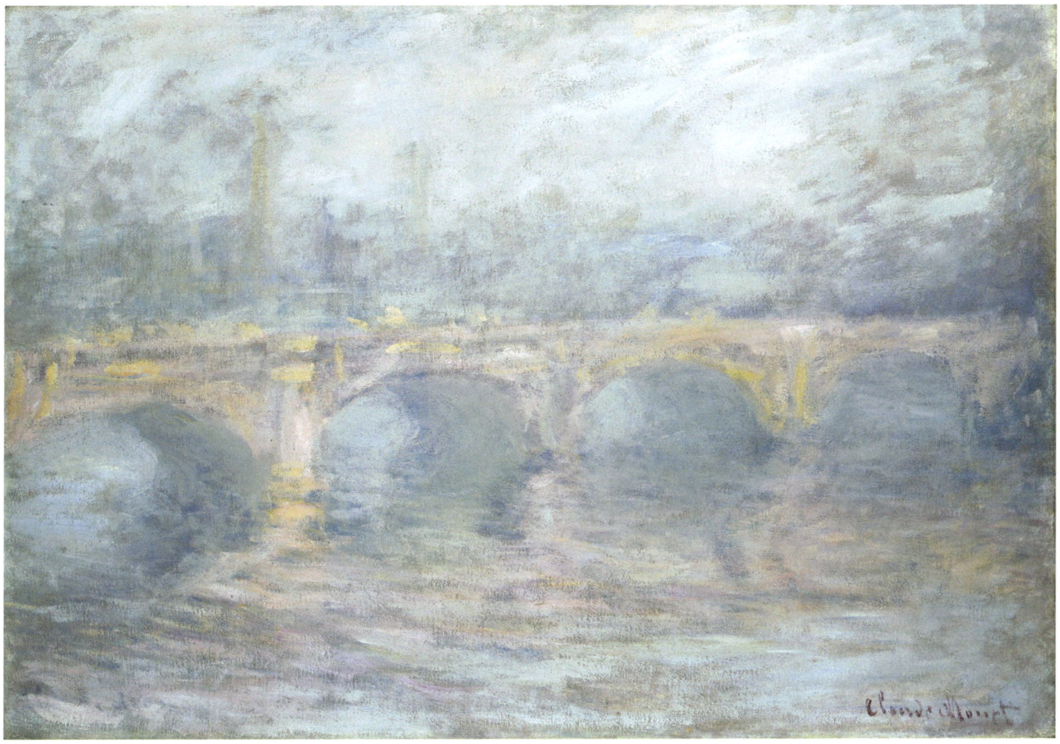

B

As the adapted signals continue along the visual pathway, signal compression occurs. Suppose a light aimed at the eye increases in even increments of radiance. The visual signals, though increasing, do so in diminishing increments. That is, the visual response is compressed compared with the radiance of the incident light. This effect can be demonstrated by intermixing colorants, as first described by Josef Albers using red and black inks (Albers 1963, plate XX-1). The results of intermixing pyrrole red and carbon black paints are shown in figure 2.6. Reducing the concentration of red while increasing the concentration of black in even increments (fig. 2.6a) produces a scale with unequal visual spacing, particularly noticeable at the two ends of the scale. Using cube-root increments of black instead (fig. 2.6b) dramatically improves visual uniformity. This fundamental property of the visual system was first discovered in the mid-nineteenth century by Ernst Weber and Gustav Fechner and was later generalized by S. S. Stevens (1961).

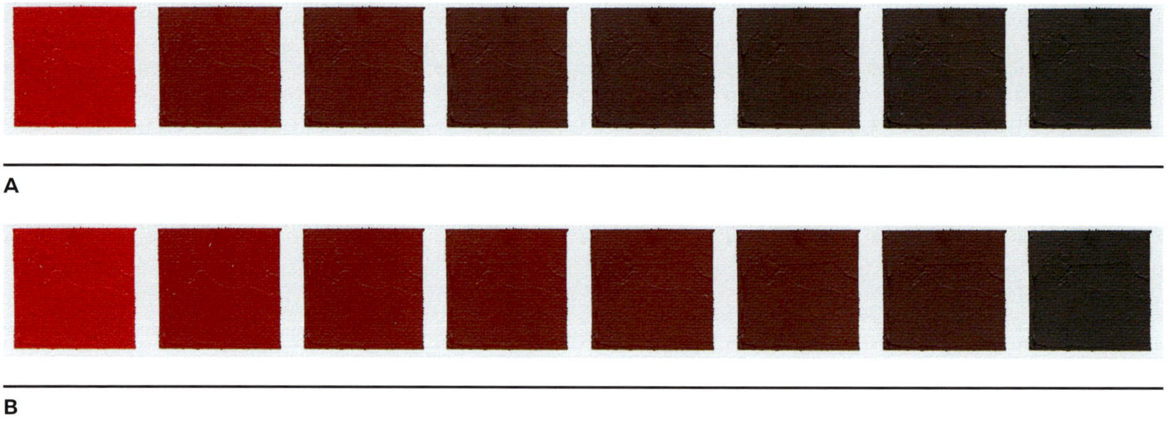

FIGURE 2.6. Mixtures of pyrrole red and carbon black in either (A) even or (B) cube-root increments by weight. The uneven increments of black result in a more uniform perceptual scale.

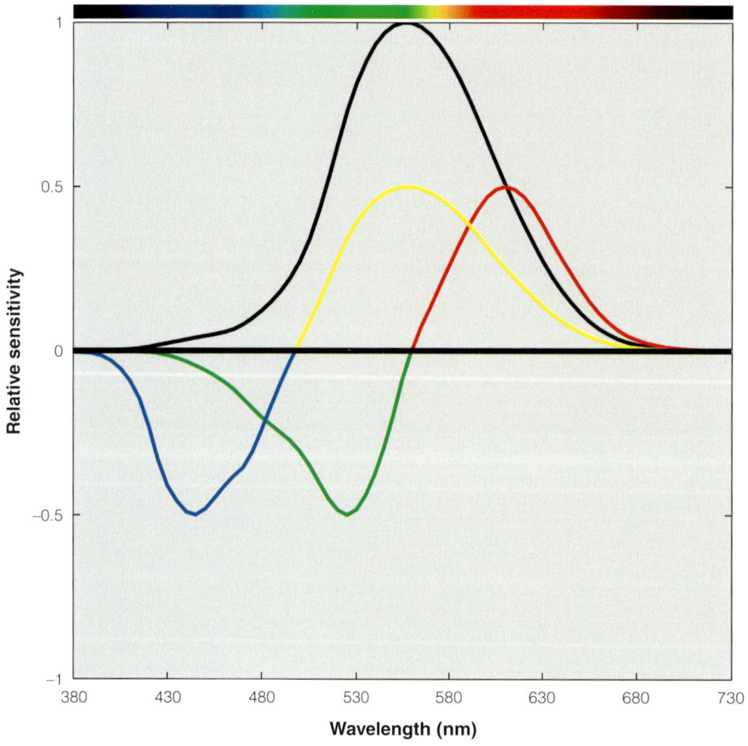

FIGURE 2.7. Spectral sensitivities of the human visual system's opponent channels.

It seems logical that our trichromatic signals, following chromatic adaptation and compression, finally send their individual signals to our brain. Thinking about our visual system as a color camera exemplifies trichromatic theory. A red color has a large L and small M and S responses; a blue color has a large S and small M and L responses; and so on. The signals are all positive, as is typical for sensors. But there are many visual phenomena that cannot be explained using trichromatic theory. For one thing, a color cannot appear simultaneously red and green or yellow and blue; that is, there is no such color as reddish green or bluish yellow. For another, people with color-vision deficiencies often confuse reds and greens.

Such observations led to a different theory of color vision, first postulated by Ewald Hering in 1878. Rather than cone signals that are all positive, the three signals sent to the brain are positive and negative: black opposed to white, red opposed to green, and yellow opposed to blue. The difficulty in accepting opponency over trichromacy is envisioning a sensor that could produce both positive and negative signals. Several vision scientists of the late nineteenth and early twentieth century proposed a two-stage theory of color vision, the first trichromatic and the second opponent (Kries 1878; Judd 1966), which was verified experimentally during the 1950s (Svaetichin 1956; Hurvich and Jameson 1957; DeValois et al. 1958).[3] The opponent stage leads to positive and negative signals. Although two-stage theory is a simplification (Daw 2012), its use is reasonable, as it explains many visual phenomena, and it serves as the basis for numerical color specification.

Subtracting the M cone response from the L cone response produces a red-opposed-to-green signal. Subtracting the M and L responses from the S response produces a yellow-opposed-to-blue signal. These are the chromatic opponent channels. Adding the M and L responses produces the black-opposed-to-white achromatic channel. As seen in figure 2.7, these arithmetic operations result in positive and negative spectral sensitivities for the chromatic channels and a positive spectral sensitivity for the achromatic channel. Two-stage color vision theory is demonstrated in figure 2.8 through the visualization of Henri Rousseau's *A Centennial of Independence* as an opponent image.

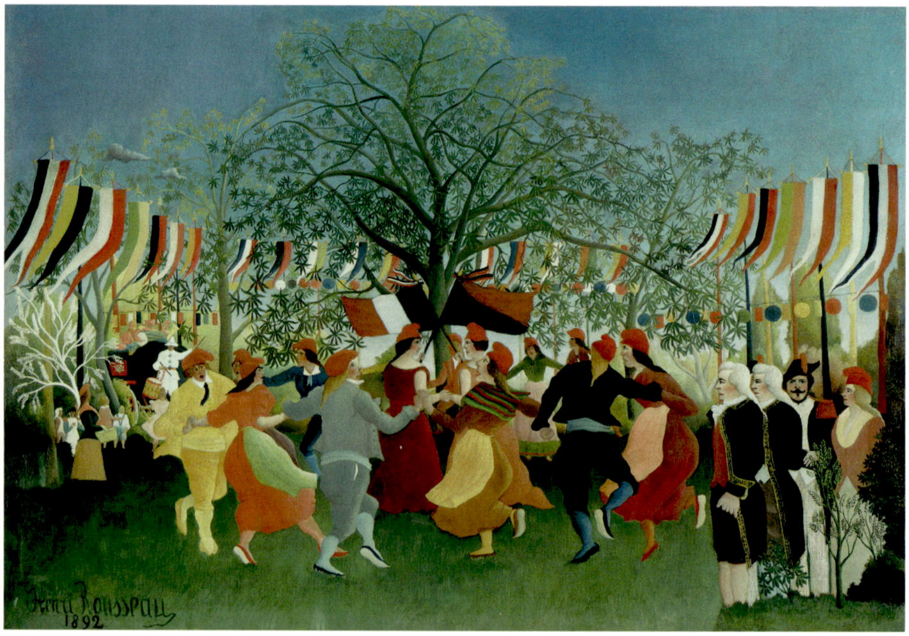

A

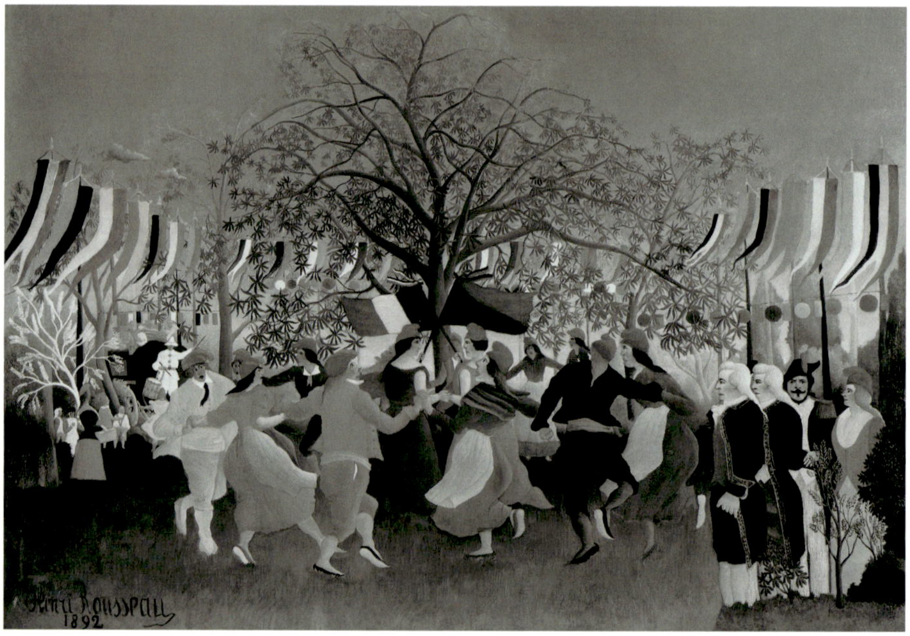

B

FIGURE 2.8.
(A) Henri Rousseau (French, 1844–1910), *A Centennial of Independence*, 1892. Oil on canvas, 111.8 × 157.2 cm (44 × 61⅞ in.). Los Angeles, J. Paul Getty Museum, 88.PA.58. The other versions show the opponent images: (B) the black-opposed-to-white channel, (C) the red-opposed-to-green channel, and (D) the yellow-opposed-to-blue channel.

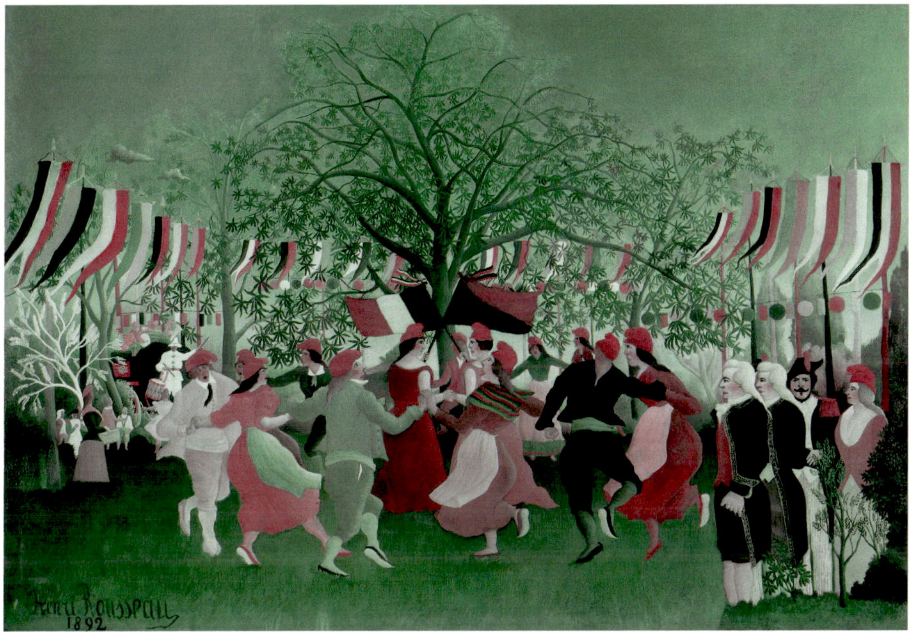

C

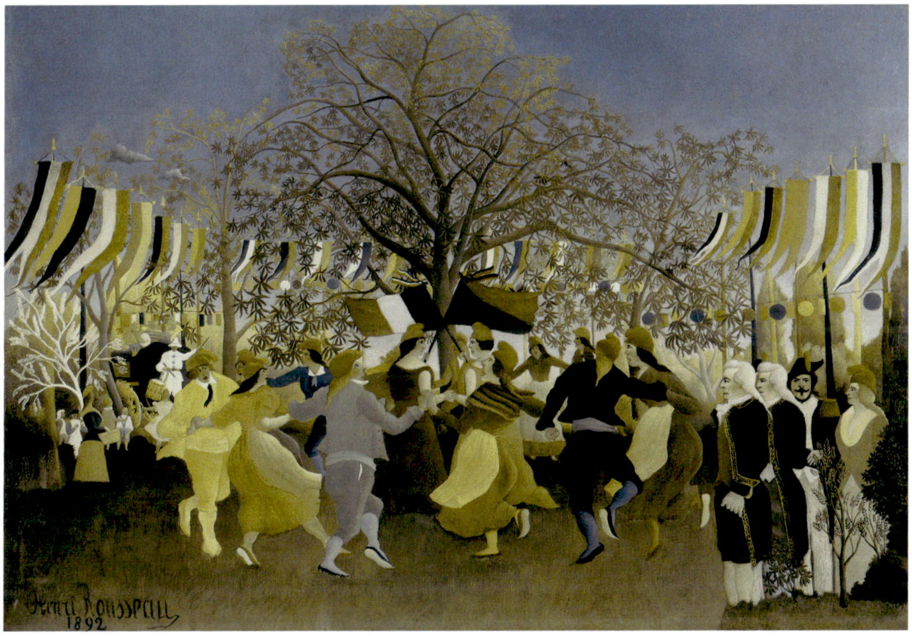

D

Variability in Color Vision

Many experiments performed over the years have revealed large differences in physiology among us. Some have involved matching individual wavelengths with three different wavelengths (for example, matching 500 nm with 450 nm, 550 nm, and 650 nm). Others have been anatomical, where the absorption and scattering properties of different optical elements were measured directly. Still others have used imaging to map our physiology—to learn, for example, how rods and cones are distributed throughout the eye (Hofer et al. 2005). Finally, gene sequencing has been used to identify receptor differences. Physiological differences in cone-receptor spectral sensitivity and changes in our lens as we age have the most profound effects in changing color vision.

As we age, optical changes in the eye's lens and in visual processing result in our vision becoming blurrier and less able to discriminate colors (Knoblauch et al. 1987; Spear 1993). A clouding of the lens is termed a cataract, and surgery is frequently performed to replace the lens with a polymeric material. This clouding is simulated for *Irises* in figure 2.9. Someone with cataracts sees the world through a blurry and yellowish lens. The yellow color is compensated partially by chromatic adaptation, and because the color change occurs over many years, the yellowness is less apparent than the lens color would indicate. Immediately after surgery, sharpness is restored; the long-term effects of the yellowing, however, result in the painting appearing bluish. With time, the blueness disappears and the painting's true colors appear.

Monet developed severe cataracts late in life, probably a result of painting mostly outdoors, *en plein air*. Following the removal of the lens of his right eye, there was a large color imbalance between his two eyes; the world seen through his right eye appeared excessively blue (Werner 1998; Marmor 2006). When he began using his right eye for painting because of the improved sharpness, he used various colored glasses to counteract the blueness. His final glasses, two years after surgery, were untinted. The length of time it took Monet to achieve a neutral chromatic adaptation is unusual; typically, it takes several months.

The spectral sensitivities of the cone receptors graphed in figure 2.3 represent the population average of color-normal observers. Individual measurements reveal changes in the location of the peak sensitivities of the L and M cone receptors (Sharpe et al. 1998). When the shifts are large, people are considered to have anomalous trichromatism, known as protanomaly and deuteranomaly for anomalies in the L and M cones, respectively. These people have color vision, but their ability to discriminate differences in hue for red and green colors is reduced. In others, one cone type is absent, resulting in dichromatic vision. People missing the L, M, or S cones have protanopia, deuteranopia, or tritanopia, respectively. Dichromatic vision is simulated[4] for Vincent van Gogh's *Irises* in figure 2.10. When

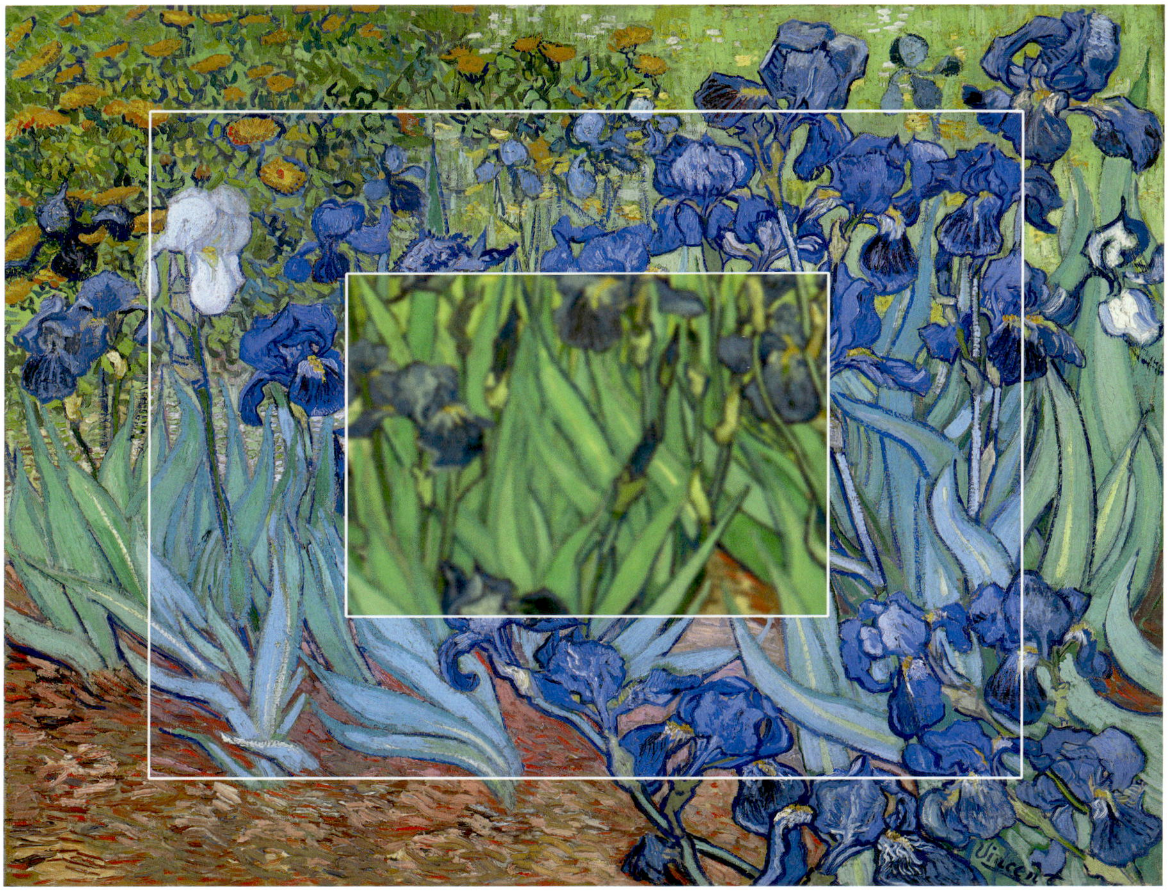

FIGURE 2.9. The demarcated areas show how Van Gogh's *Irises* would look if seen through cataracts (center), immediately after lens replacement (middle), and several months later (outer frame).

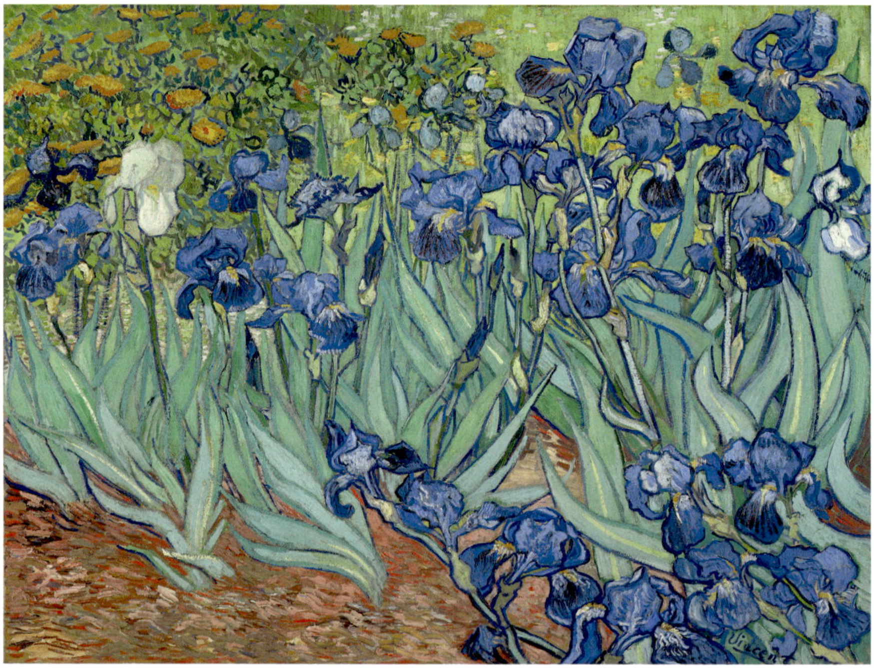

A

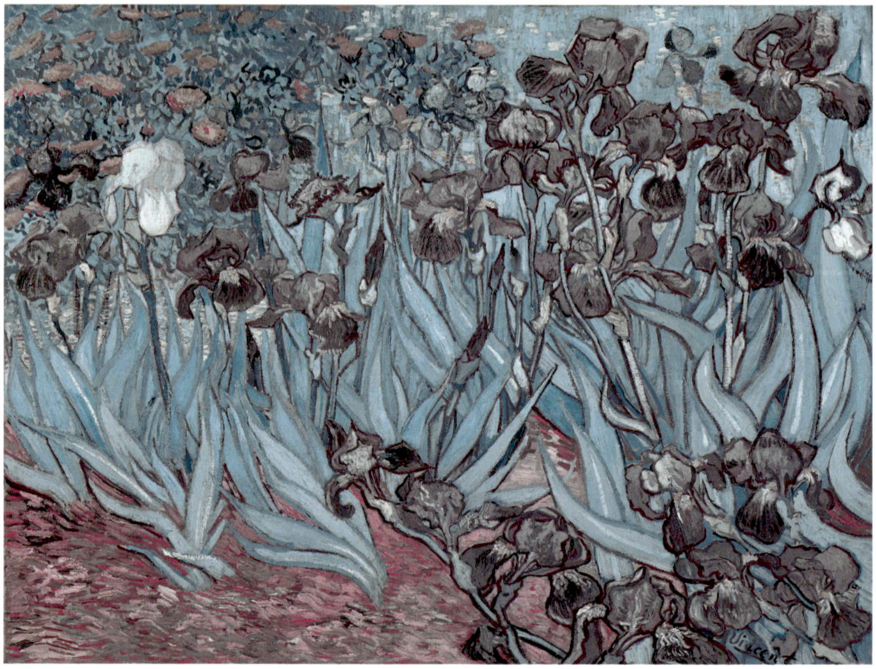

B

FIGURE 2.10.
(A) Vincent van Gogh (Dutch, 1853–1890), *Irises,* 1889. Oil on canvas, 71.1 × 93 cm (28 × 36⅝ in.). Los Angeles, J. Paul Getty Museum, 90.PA.20. The other versions simulate the painting's appearance for people missing (B) S cones, (C) M cones, and (D) L cones, conditions known as tritanopia, deuteranopia, and protanopia, respectively. Notice that the reddish soil is darker for protanopes (D) than for deuteranopes (C).

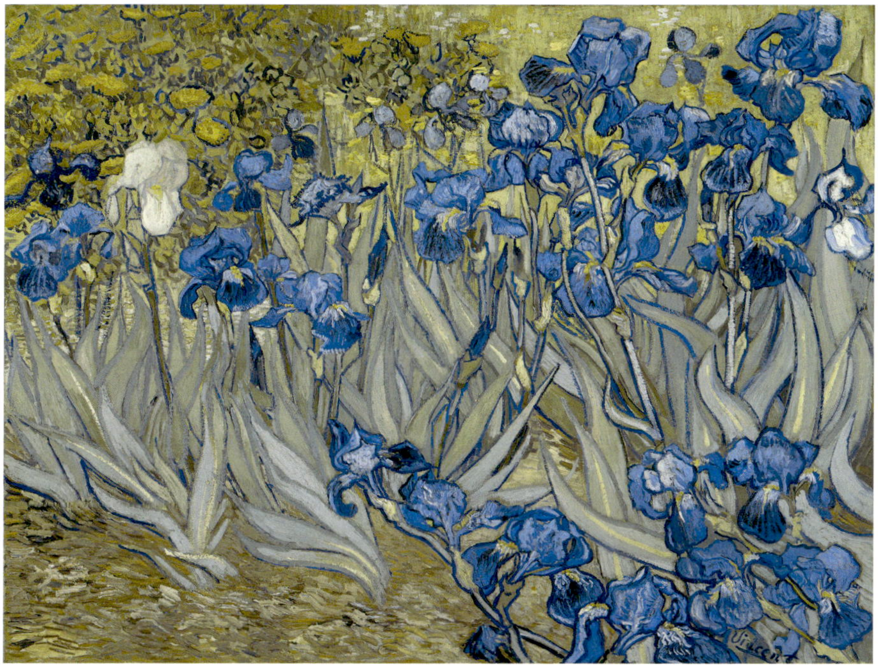

C

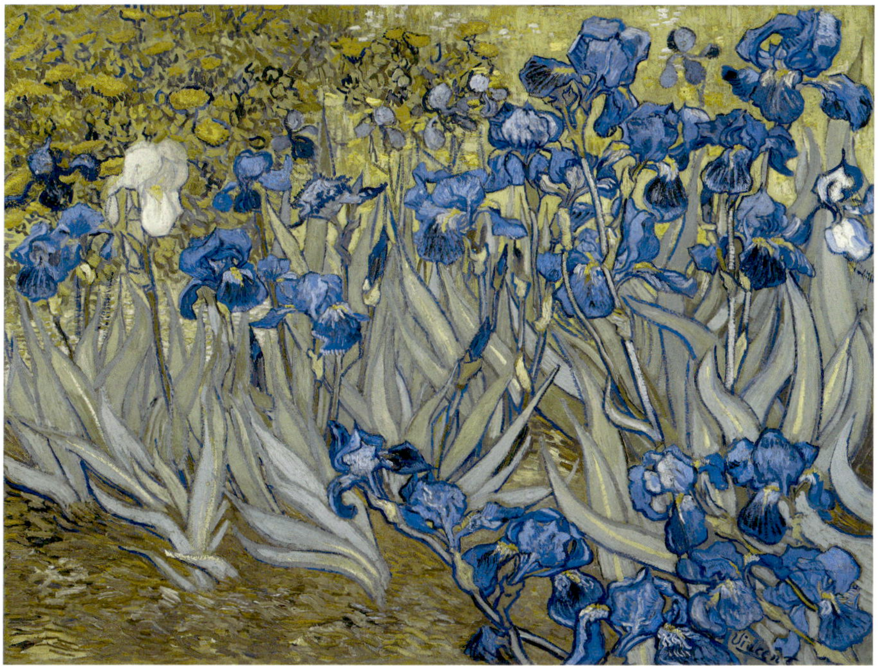

D

L or M cones are missing, the red-opposed-to-green opponent channel is disrupted, resulting in a world of approximate yellows, blues, and neutrals. The absence of S cones disrupts the yellow-opposed-to-blue opponent channel, resulting in a world of approximate reds, greens, and neutrals. Color deficiencies are found in 8% of the male and 0.5% of the female population; a variety of screening tests are available for detection.

Spatial Vision

De la loi du contraste simultané des couleurs et de l'assortiment des objets colorés (The Principles of Harmony and Contrast of Colors), written in 1839 by the French chemist Michel Eugène Chevreul, was the first book to describe visual illusions caused by the juxtaposition of colors. Systematic studies of tapestries led to his law of simultaneous contrast: "In the case where the eye sees at the same time two contiguous colours, they will appear as dissimilar as possible, both in their optical composition and in the strength of their colour" (Chevreul 1839). An example similar to one in his book is shown in figure 2.11, where six stars with identical spectral reflectance each have a unique color because each is surrounded by a different colored background.[5] This illusion is caused by connections between cone receptors such that signals from neighboring receptors affect a specific receptor. Using opponent colors maximizes simultaneous contrast effects.

Another seminal book demonstrating simultaneous contrast is Josef Albers's *Interaction of Color* (Albers 1963). In the book, based on student work, the effects of surround color are exquisitely explored.

The illusion shown in figure 2.12 changes with viewing distance.[6] Up close, simultaneous contrast is seen; at distance, neighboring colors combine spatially, known as assimilation. Both effects are predicted by opponent color vision theory. The dramatic illusion shown in figure 2.13 reveals the complexity of these interconnections, seemingly defying explanation.[7]

Understanding spatial vision begins by considering imagery in terms of its detail. The image in figure 2.14 has alternating red and green sinusoidal stripes whose widths become narrower from left to right. The number of stripes over a defined length is known as spatial frequency. The wide stripes have low spatial frequency, while the narrow stripes have

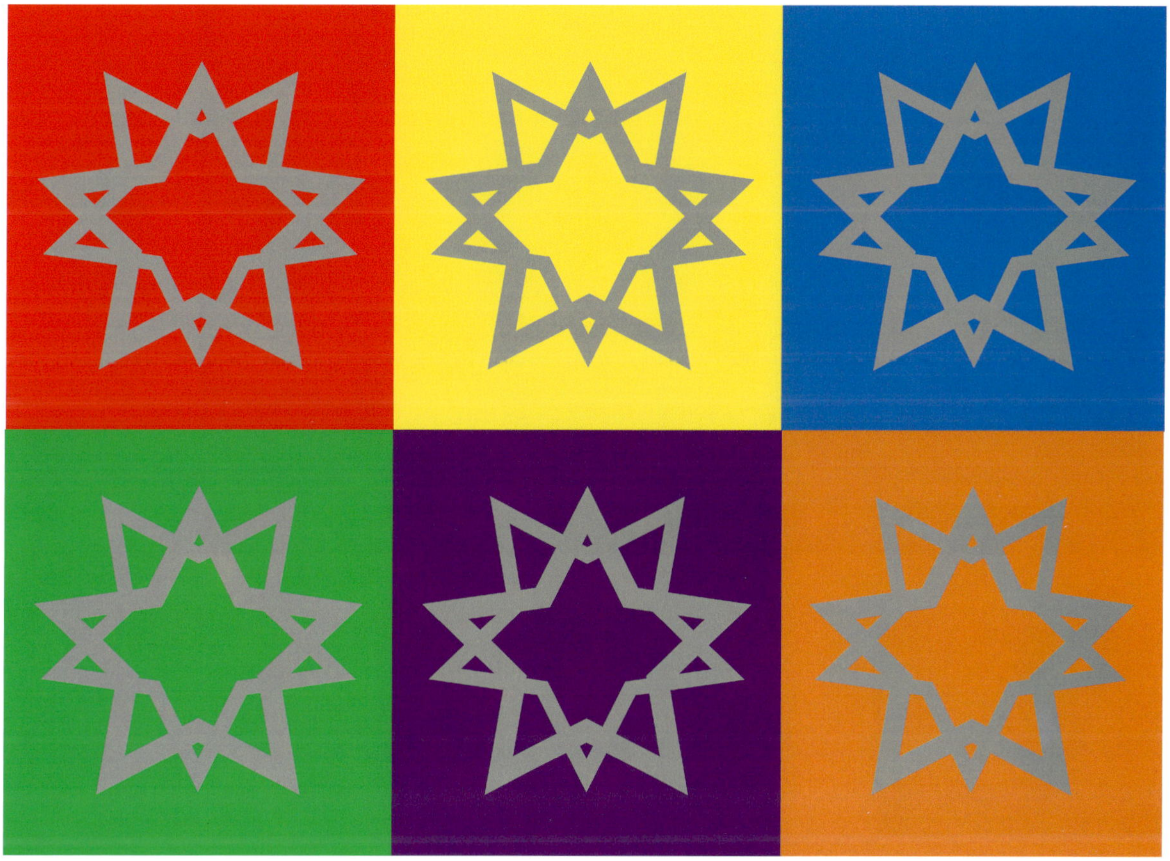

FIGURE 2.11. The same—that is, physically identical—gray color shown against different colored backgrounds. The appearance of the gray is affected by the background color.

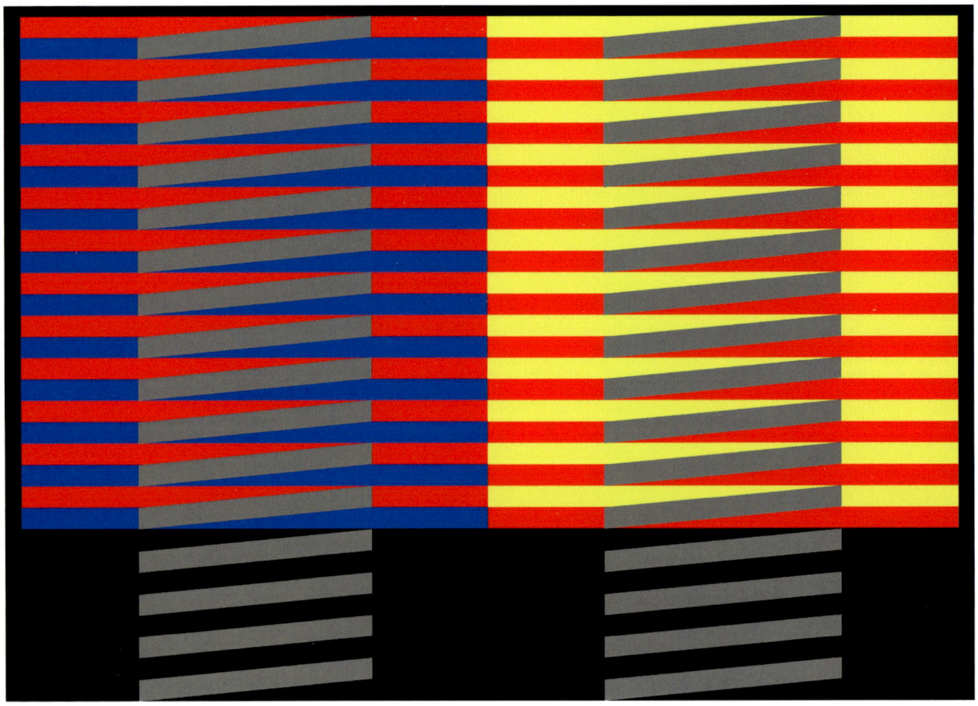

A

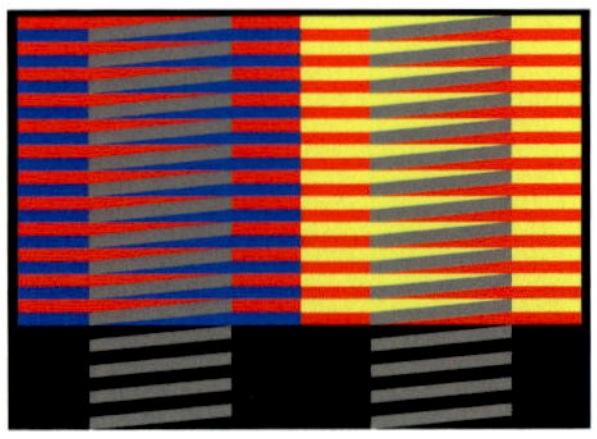

B

FIGURE 2.12. The gray stripes are all the same physical color, as are the red stripes, but because they are surrounded by different colors, they do not appear to be the same. The gray stripes appear shaded; the hue of the red stripes changes dramatically depending on image size, viewing distance, or whether the red stripes are surrounded by blue or yellow stripes. (A) In the large image, the effect is complementary; this means that, for instance, the red stripes appear more yellowish when surrounded by the blue stripes. (B) In the small image (but also when viewing the large image from a distance), the effect is additive, which means that the red stripes appear more bluish when surrounded by the blue stripes. These effects are sometimes referred to as simultaneous contrast and assimilation, respectively.

high spatial frequency. For the three panels in figure 2.15, the striped image was subdivided into three different spatial-frequency regions: low, medium, and high.[8] In the low-frequency image, the widest stripes are still evident, while the rest of the image is blurry and yellowish in color. The medium-frequency image has a greater range of distinct stripes. The high-frequency image has distinct narrow stripes and blurred wide stripes. When defining spatial properties of our visual system, field of view is used rather than physical length. An object seen up close extends across a larger field of view than the same object viewed from a distance. Moving away from the object increases its spatial frequency. Our ability to discern different spatial frequencies for a given field of view is known as acuity. In figure 2.16, Van Gogh's *Irises* was analyzed spatially by separating a detail image of the painting into three ranges of spatial frequency. Low spatial frequencies are areas of uniform color, while high spatial frequencies reveal edge detail and specular highlights. The medium-spatial-frequency image appears as if lit very diffusely or viewed from a greater distance.

Opponent vision results from different cone types adding to or subtracting from one another. This can only occur spatially, since our receptors are arranged as a mosaic. The spatial processing results in different acuity for the three opponent channels, where our black-opposed-to-white channel has the highest acuity and the yellow-opposed-to-blue channel has the lowest acuity. Looking at the achromatic channel, we can discern a much greater range of spatial frequencies than in either of the chromatic channels. To create figure 2.17, an image of *Irises* was transformed to an opponent-vision representation and divided into achromatic and chromatic images.[9] Each of the images contains complete spatial information, yet the detail is clearer in the achromatic channel. Although there is a range of values in the chromatic image, we cannot see these differences because of low acuity. The range can be seen only by turning the colored image into grayscale.

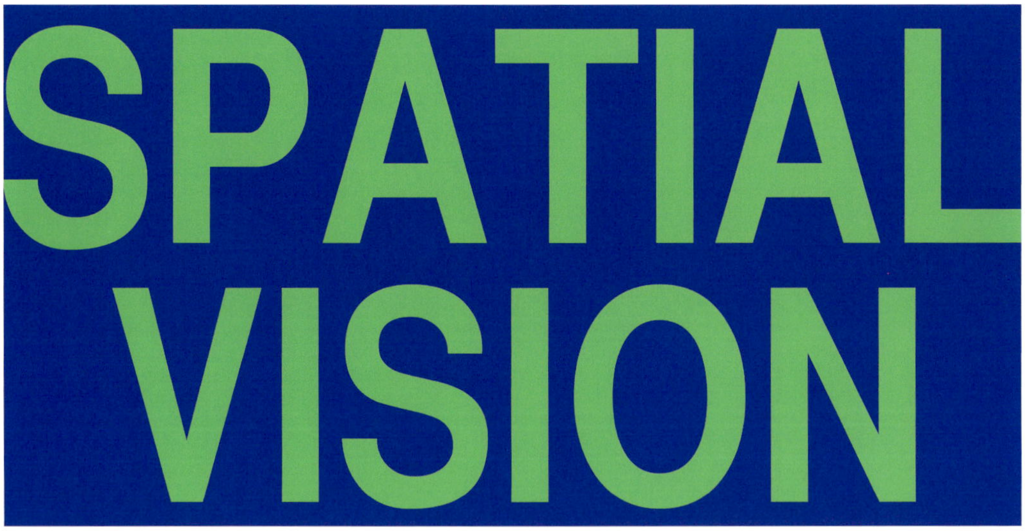

A

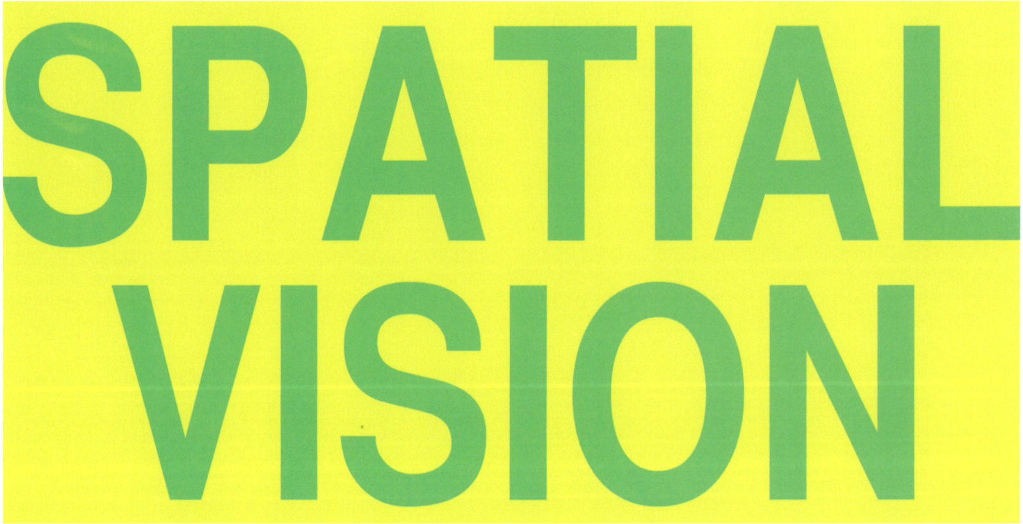

B

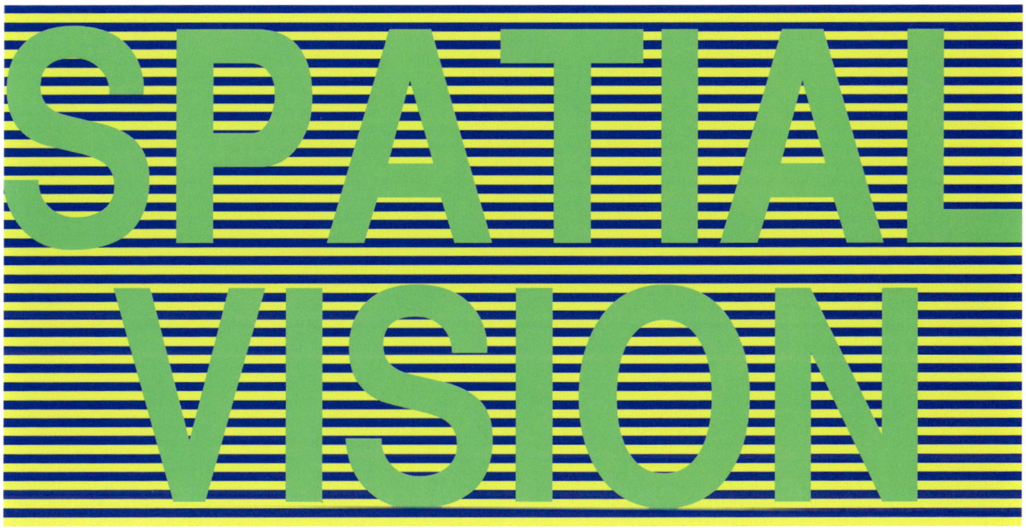

C

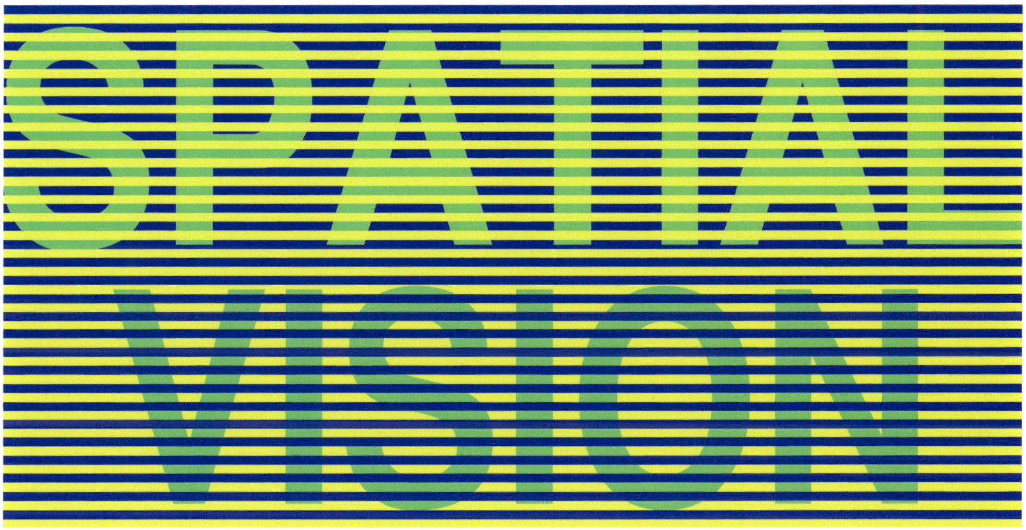

D

FIGURE 2.13. The color of the phrase "spatial vision" appears to change as the background changes from (A) purple, to (B) yellow, to (C) purple and yellow stripes. These changes can be predicted using the principles of simultaneous contrast and assimilation. But when (D) the yellow stripes are placed in front of "spatial" and the purple stripes are placed in front of "vision," there is a dramatic change in each word's color, not predicted by simultaneous contrast or assimilation, demonstrating that our spatial color vision is highly complex.

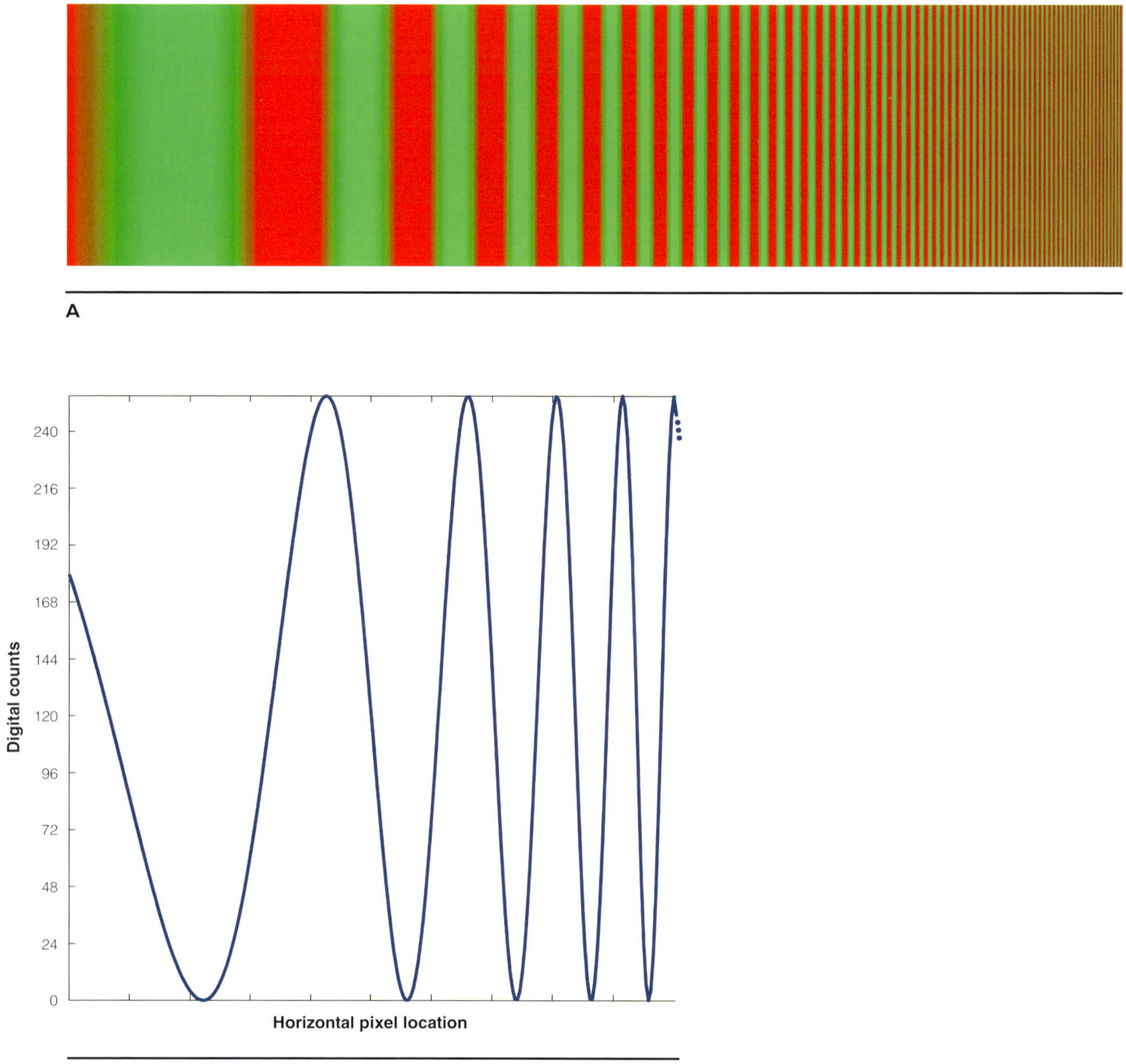

A

B

FIGURE 2.14. (A) Color image of red and green sinusoidal stripes of decreasing width (or increasing spatial frequency), moving from left to right. (B) The corresponding image data—digital counts (defined on p. 166)—are plotted for the red channel. Thus the spatial frequency of the sine wave is increasing.

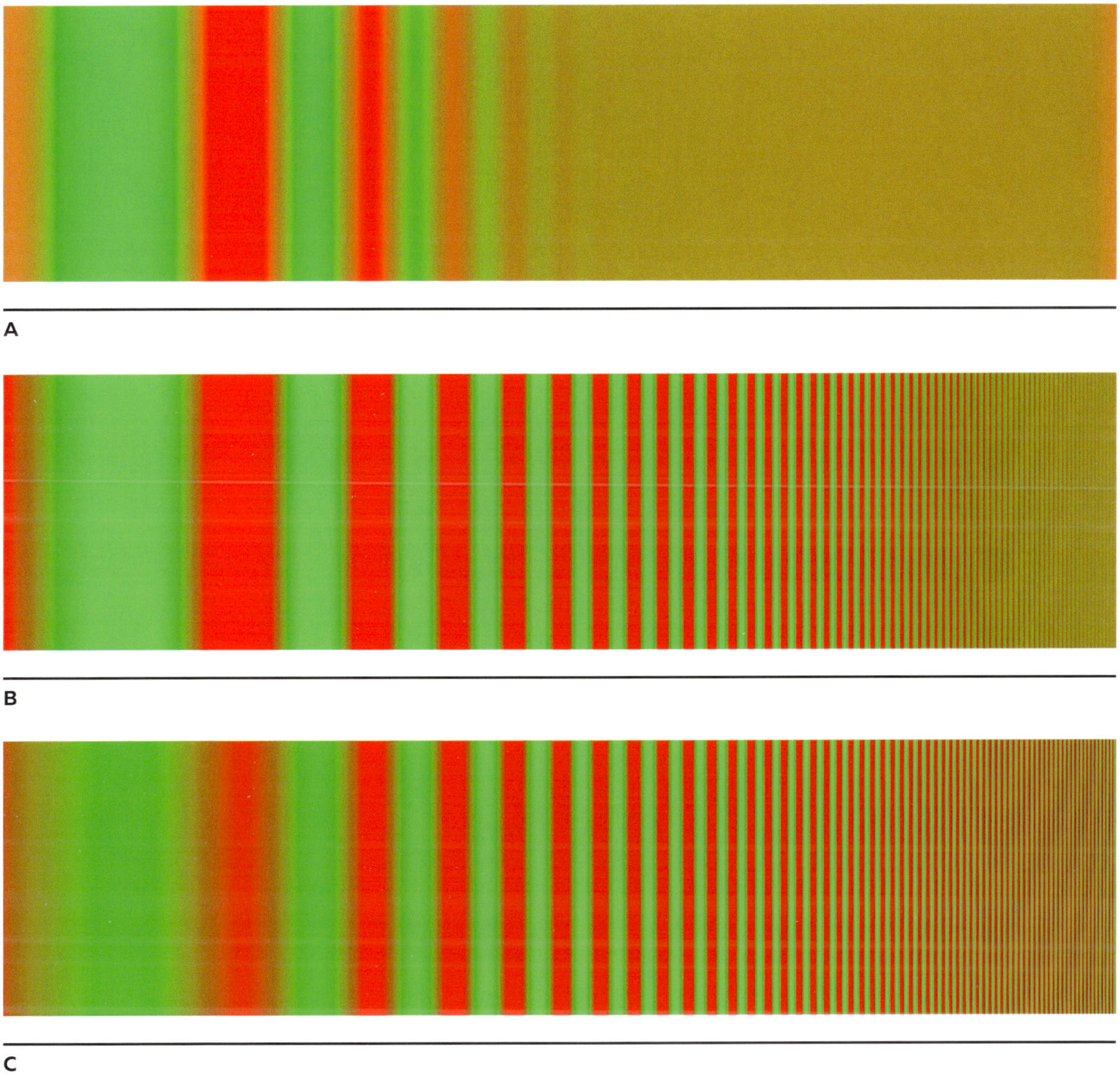

FIGURE 2.15. Dividing the sinusoidal stripes of figure 2.14 into (A) low, (B) medium, and (C) high spatial frequencies.

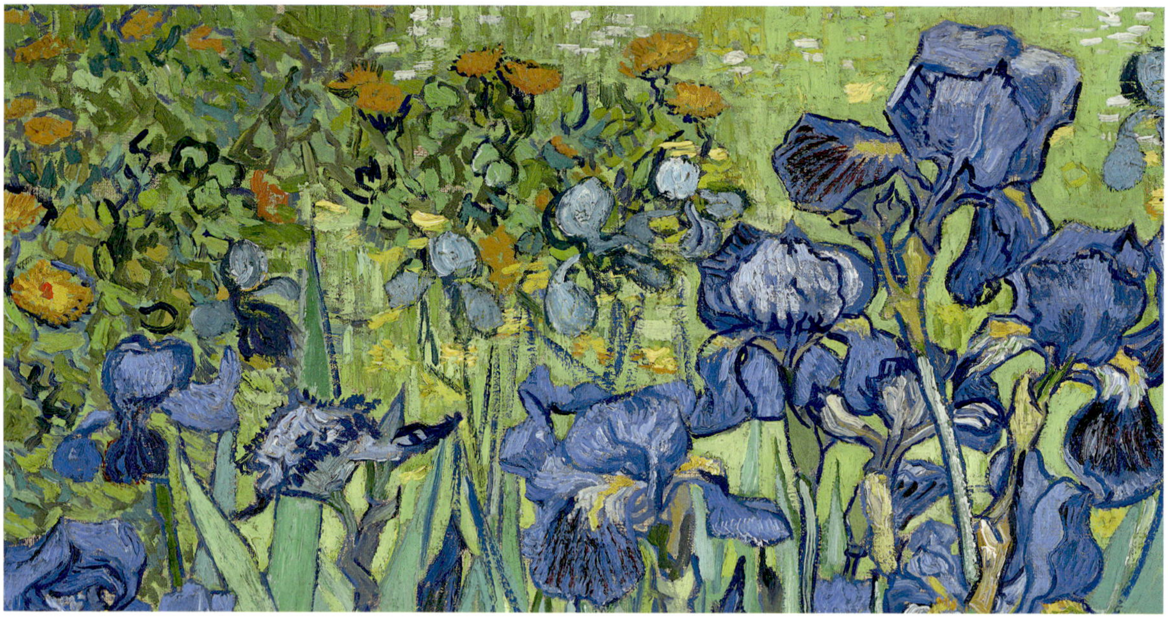

A

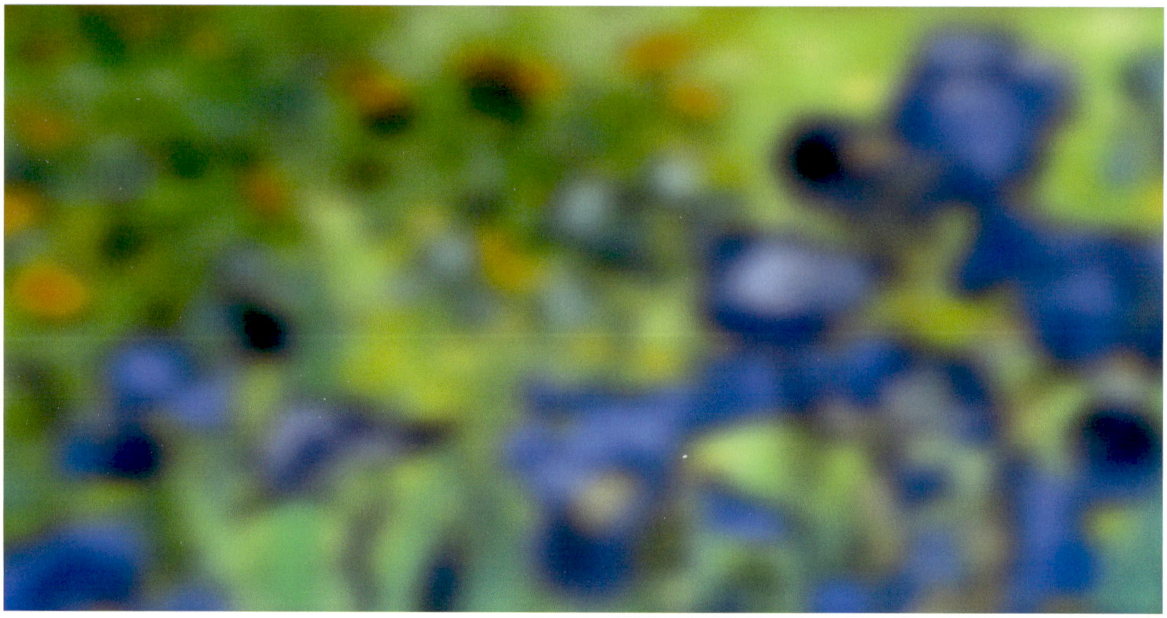

B

FIGURE 2.16. Detail of Van Gogh's *Irises* shown at (A) full spatial frequency and divided into (B) low, (C) medium, and (D) high spatial frequencies.

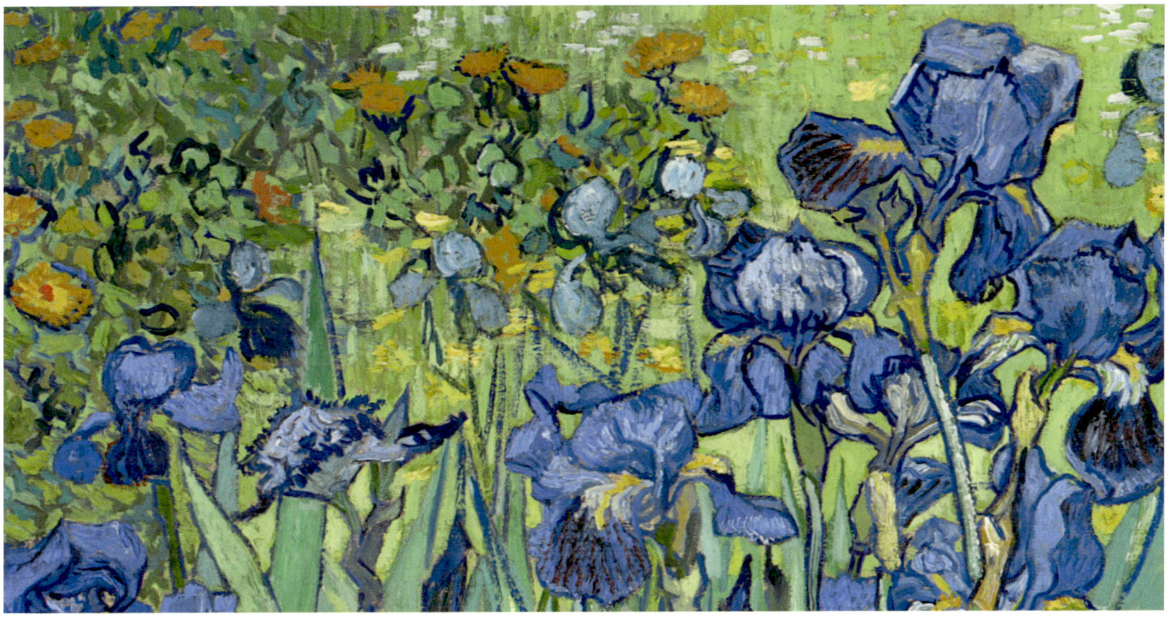

C

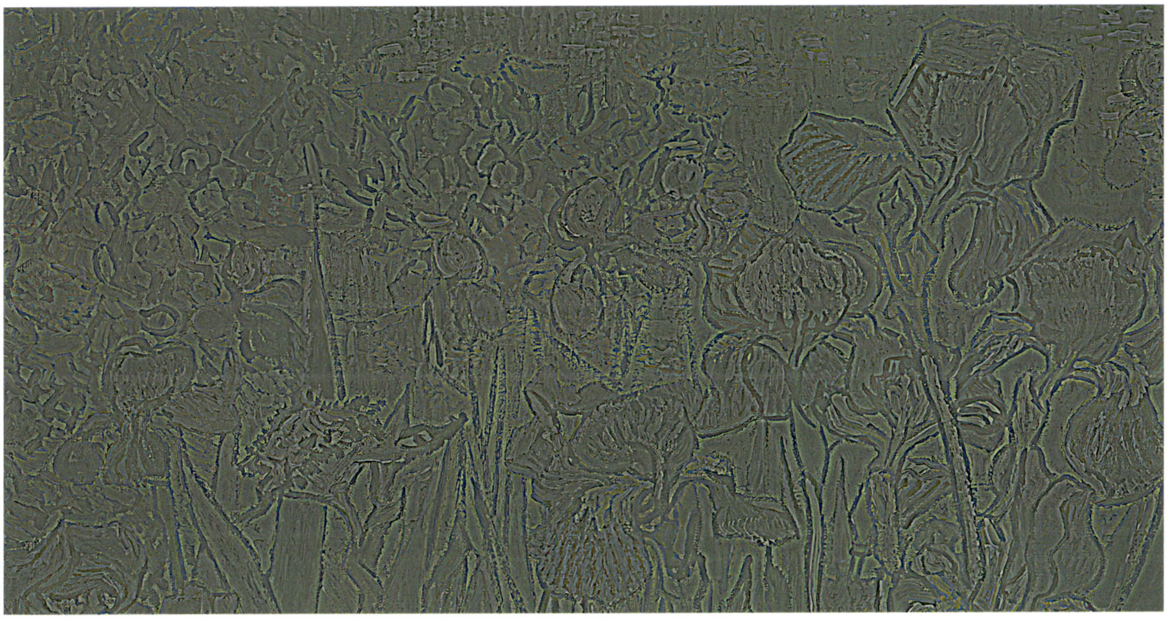

D

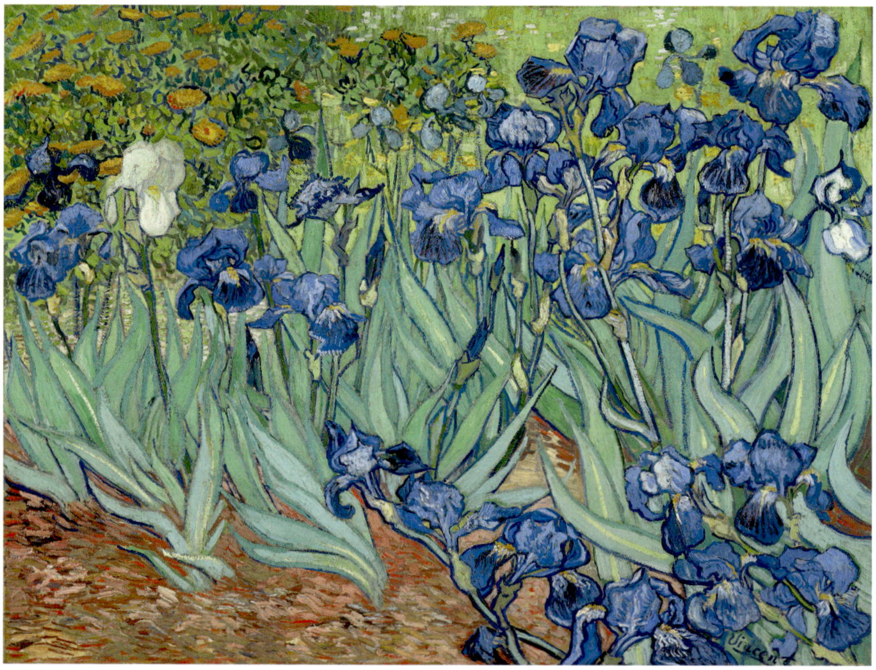

A

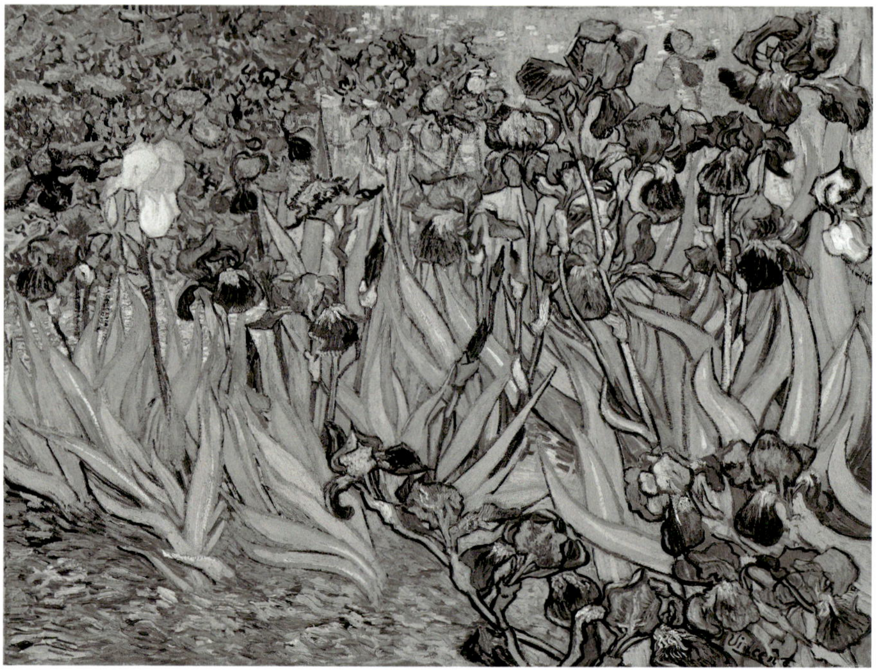

B

FIGURE 2.17.
(A) Van Gogh's *Irises* divided into (B) achromatic and (C) chromatic images. To produce (D), the chromatic image was transformed to grayscale, revealing image detail not seen in the colored image.

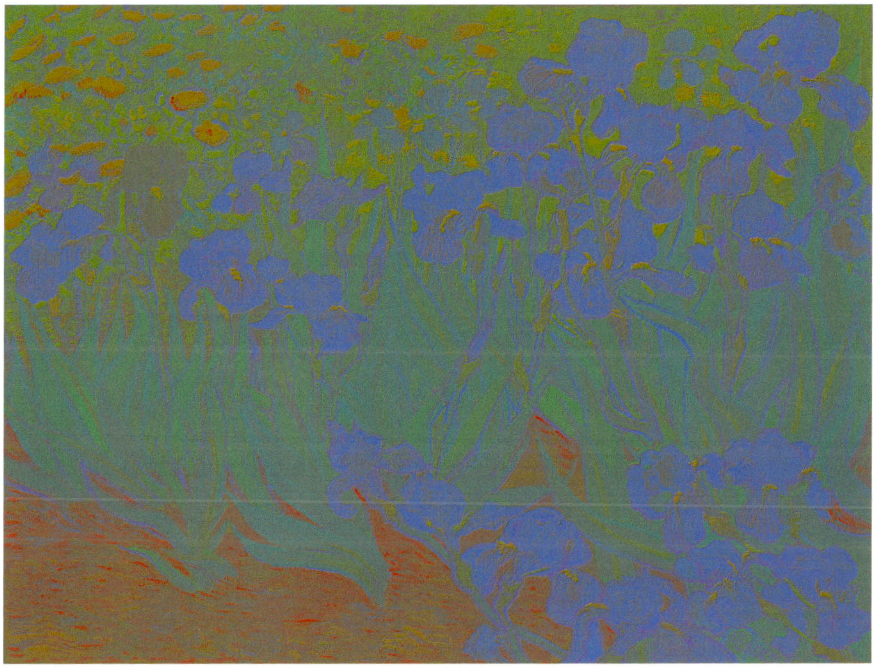

C

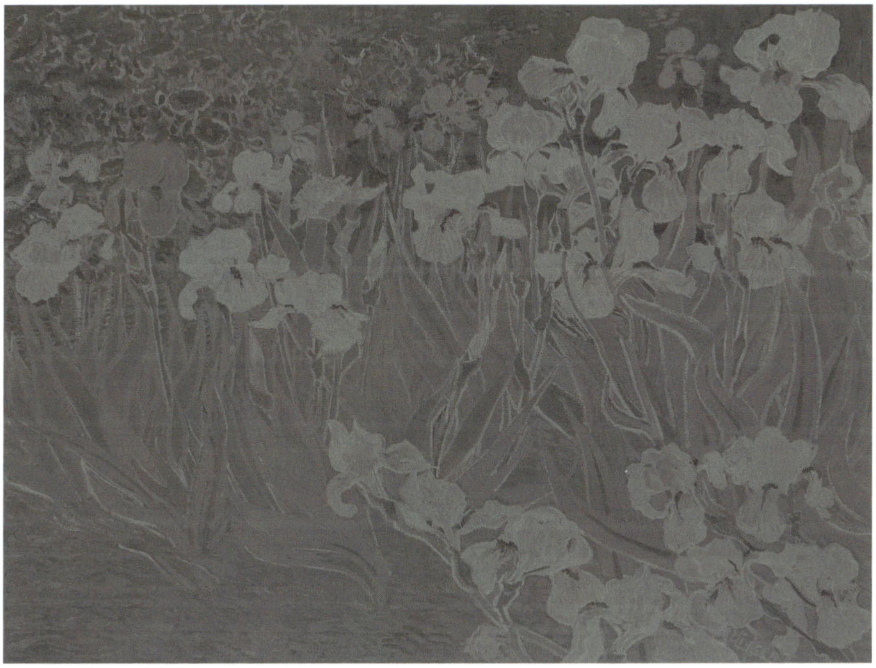

D

Another way to demonstrate differences in acuity is to blur (defocus) each opponent channel individually and see the visual effect, as shown in figure 2.18. The blurring of the achromatic image is very visible because most of the detail is lost. Blurring the chromatic channels has a negligible effect on sharpness.

Image-file compression such as JPEG is a spatially based compression where color-image data are transformed to achromatic and chromatic channels. The chromatic channels are compressed more than twice as much as the achromatic channel. Reducing image detail predominantly in the chromatic channels reduces file size without a visible degradation in image quality.

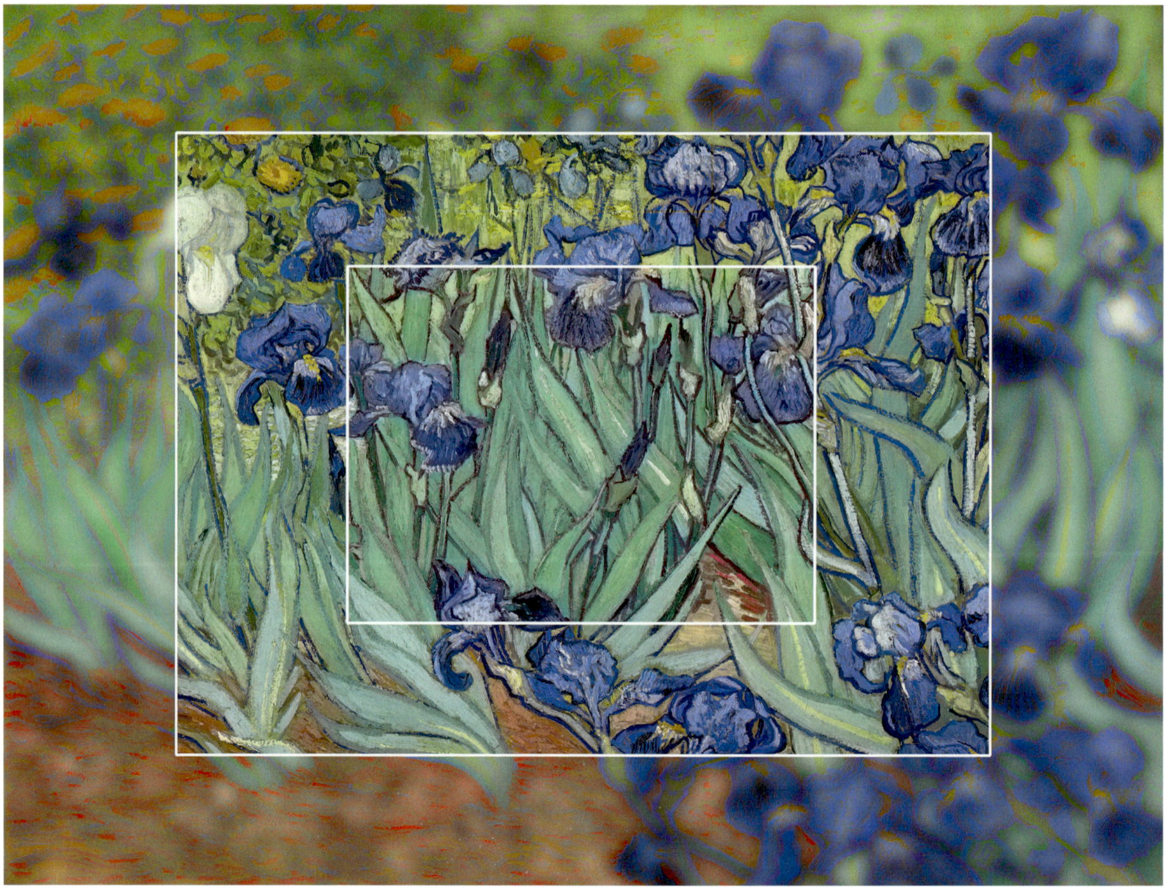

FIGURE 2.18. Van Gogh's *Irises* with an equal amount of blurring in the achromatic (outside), red-opposed-to-green (middle), and yellow-opposed-to-blue (center) channels. The blurring is most visible in the achromatic channel and least visible in the yellow-opposed-to-blue channel.

How might an artist use this knowledge? Our ability to see detail and depth depends on our achromatic channel. Artwork looks more realistic when lightness is well rendered, as shown in figure 2.19, *An Old Man in Military Costume* by Rembrandt van Rijn. The lightness channel has the same realism as the painting, appearing almost photographic. Portraits by the Russian expressionist Alexei Jawlensky, such as the one shown in figure 2.20, often contain vivid, non-skin-tone colors. When lightness is rendered correctly, the depth appears realistic.

Our ability to discern detail reduces at a distance, since the smaller field of view causes an increase in the spatial frequency of the details. Objects in the distance should have reduced detail, as in figure 2.21, Édouard Manet's *The Rue Mosnier with Flags,* where the flags become less distinct at the end of the street.

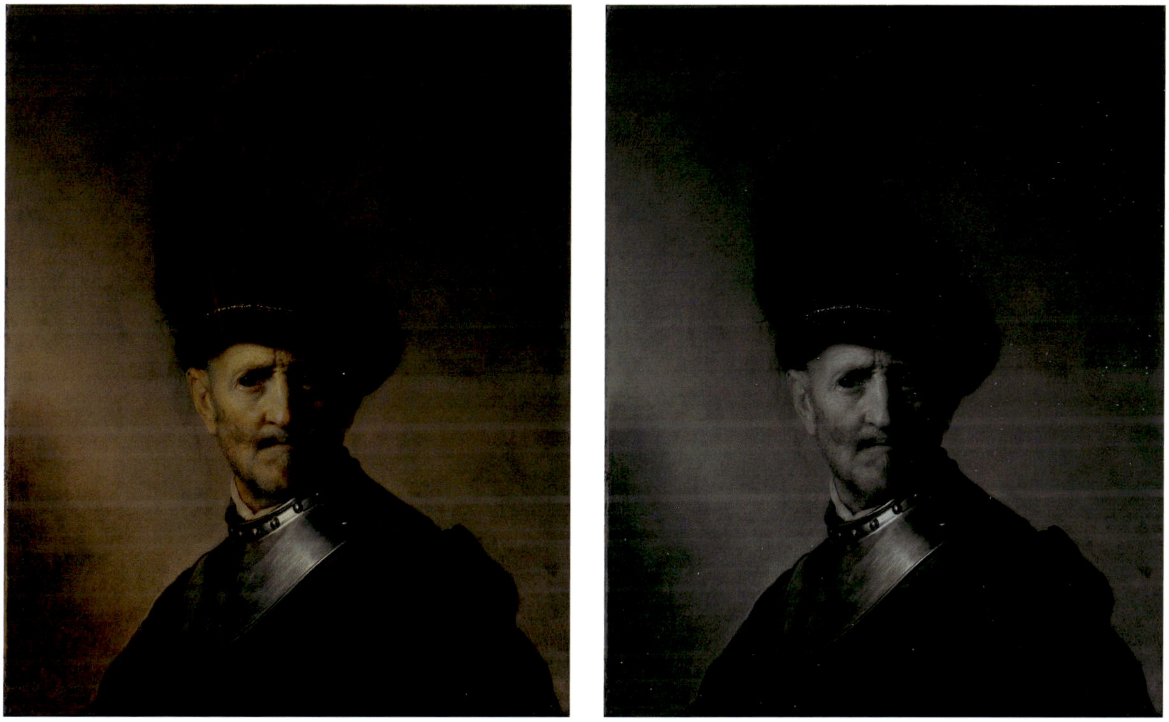

A B

FIGURE 2.19. (A) Rembrandt van Rijn (Dutch, 1606–1669), *An Old Man in Military Costume,* ca. 1630–1631. Oil on panel, 66 × 50.8 cm (26 × 20 in.). Los Angeles, J. Paul Getty Museum, 78.PB.246. (B) The image of the lightness-channel information.

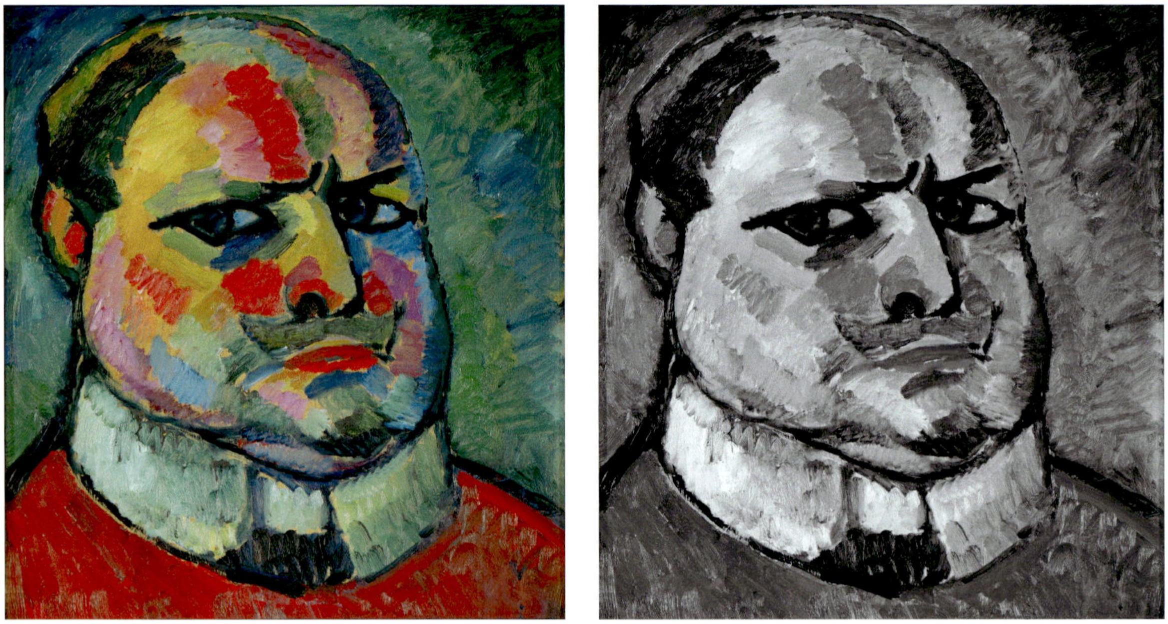

A B

FIGURE 2.20. (A) Alexei Jawlensky (Russian, 1864–1941), *Self-Portrait (Selbstbildnis)*, 1912. Oil on cardboard, 53.6 × 48.5 cm (21 × 19⅛ in.). Museum Wiesbaden. (B) The image of the lightness-channel information.

Our spatial sensitivity also reduces from the center of our visual system toward the periphery because our cones become increasingly sparse; in their place are rods. If the composition seems to draw our attention to a specific area, it may have greater image detail, as we see in figure 2.22, a pastel by Francis Cotes, where the brothers are much sharper than the background. Showing only the high spatial frequencies reveals the difference in sharpness between the brothers and the background.

Painters have used lightness contrast to overcome the limited lightness range of artist materials, particularly when the final surface is rough, as in the case of an unvarnished painting, drawing, or fresco. The detail of *An Old Man in Military Costume* in figure 2.23 reveals how Rembrandt surrounds specular highlights with dark colors to increase the lightness of the highlights, improving the metallic quality of the mantle. In a related technique illustrated in figure 2.24, Georges Seurat often lightened the background at the edge of an object on the shadow side to increase three-dimensionality, known as countershading. Increasing the

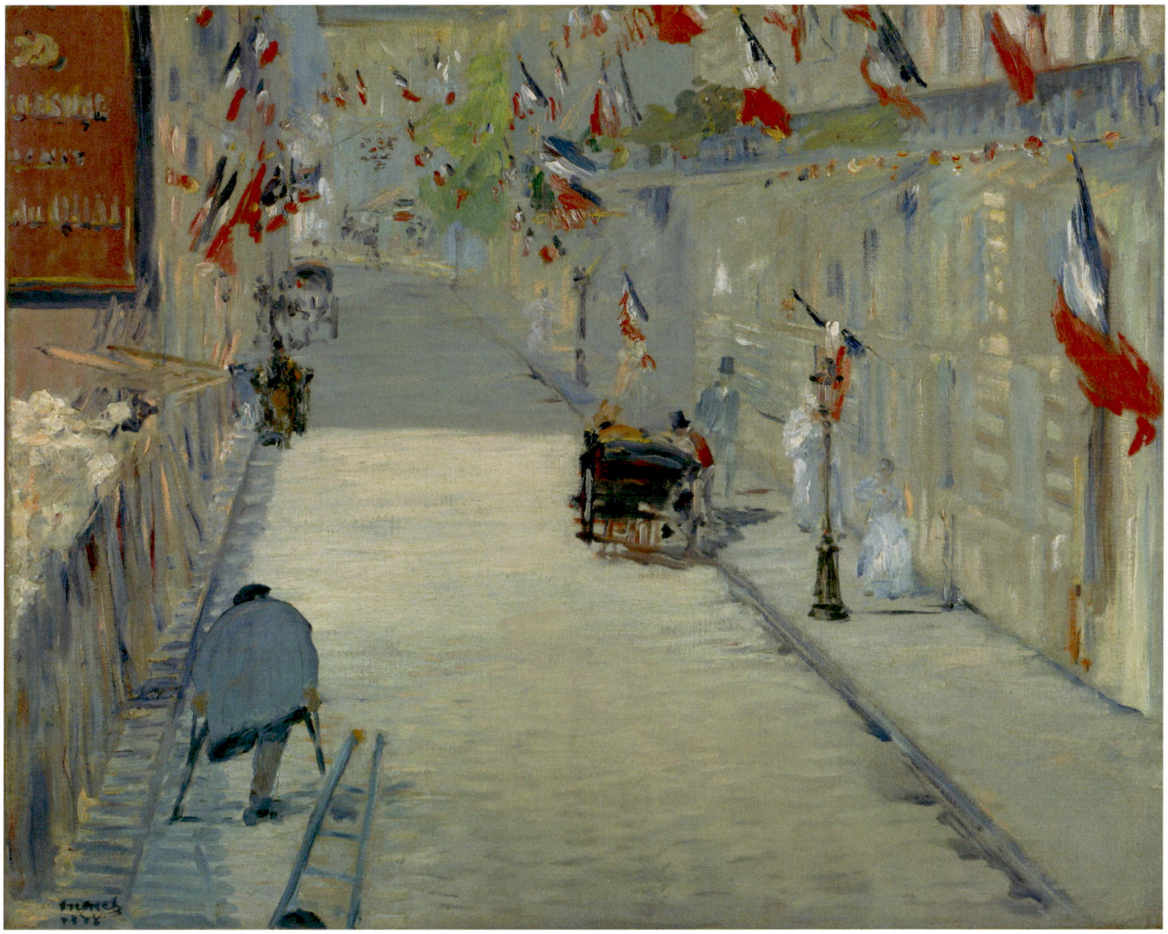

FIGURE 2.21. Édouard Manet (French, 1832–1883), *The Rue Mosnier with Flags*, 1878. Oil on canvas, 65.4 × 80.7 cm (25¾ × 31¾ in.). Los Angeles, J. Paul Getty Museum, 89.PA.71. The flags are less distinct at the end of the street.

lightness contrast darkens the edge of the object beyond what is achievable with Conté crayon alone. Countershading is particularly effective when the main figure is already a dark color.

How might an art conservator use this knowledge? When performing imitative inpainting,[10] matching the chromatic content is less important than matching the achromatic content if the area has appreciable image detail. When the area has minimal image detail, as is true for skies and color-field paintings, matching both kinds of content is critical. This point is illustrated in figure 2.25 with a return to Cotes's pastel.

FIGURE 2.22.
(A) Francis Cotes (English, 1726–1770), *Portrait of Joseph Gulston and His Brother John Gulston,* 1754. Pastel, 67.3 × 82.6 cm (26½ × 32½ in.). Los Angeles, J. Paul Getty Museum, 99.PC.33. The brothers are drawn much more sharply than the background, as is especially obvious in (B), the image with only the high spatial frequencies shown.

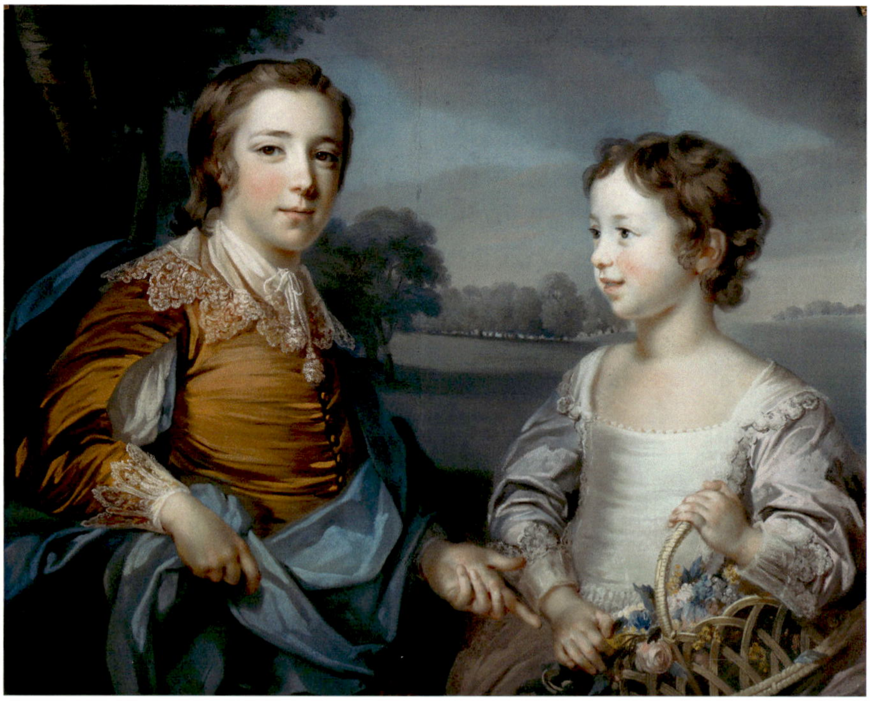

A

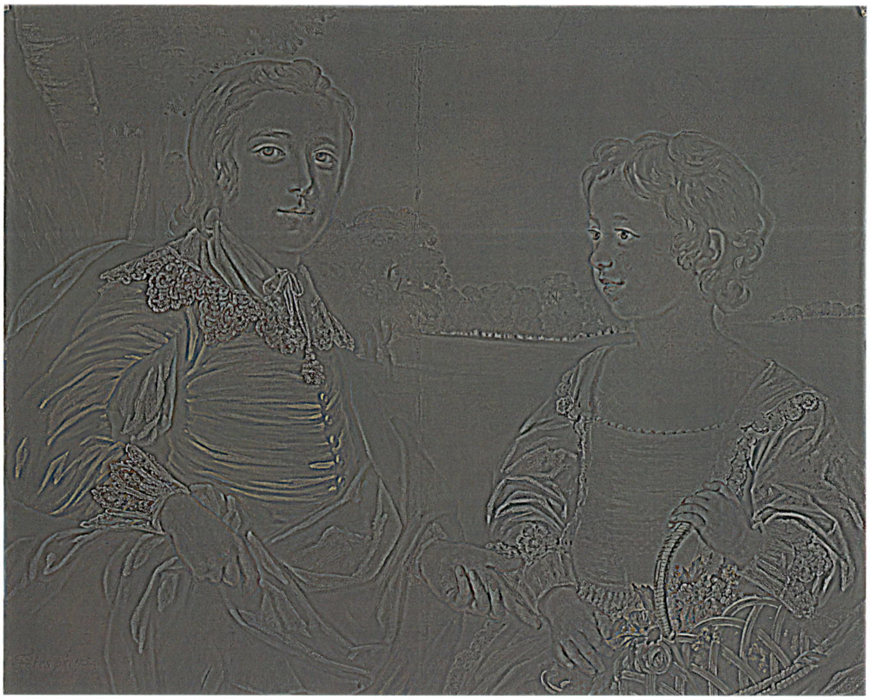

B

FIGURE 2.23. Detail of Rembrandt's *Old Man in Military Costume* (fig. 2.19a). The white passages are specular highlights, accentuated by being surrounded with dark colors.

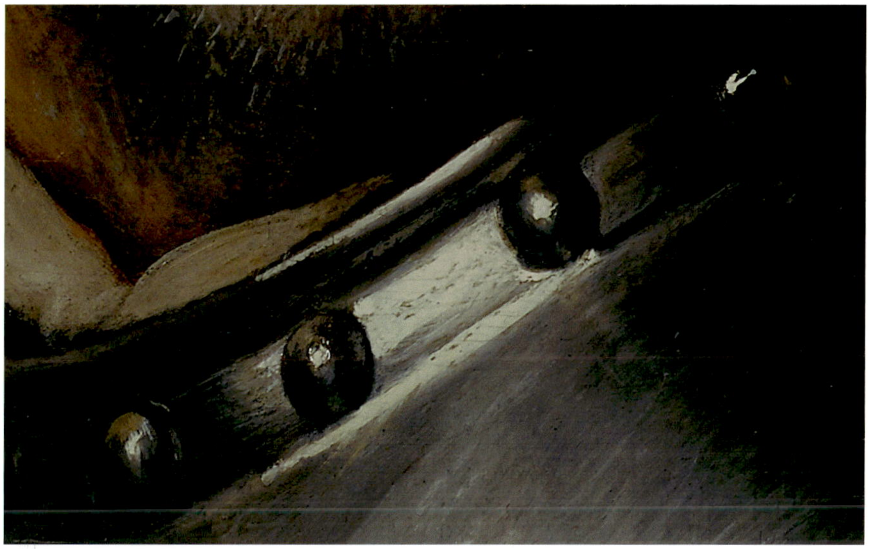

FIGURE 2.24. Georges Seurat (French, 1859–1891), *Woman Strolling (Une élégante)*, ca. 1884. Conté crayon on Michallet paper, 31.8 × 24 cm (12½ × 9½ in.). Los Angeles, J. Paul Getty Museum, 2000.30. Lightening the background behind the left side of the figure increases three-dimensionality.

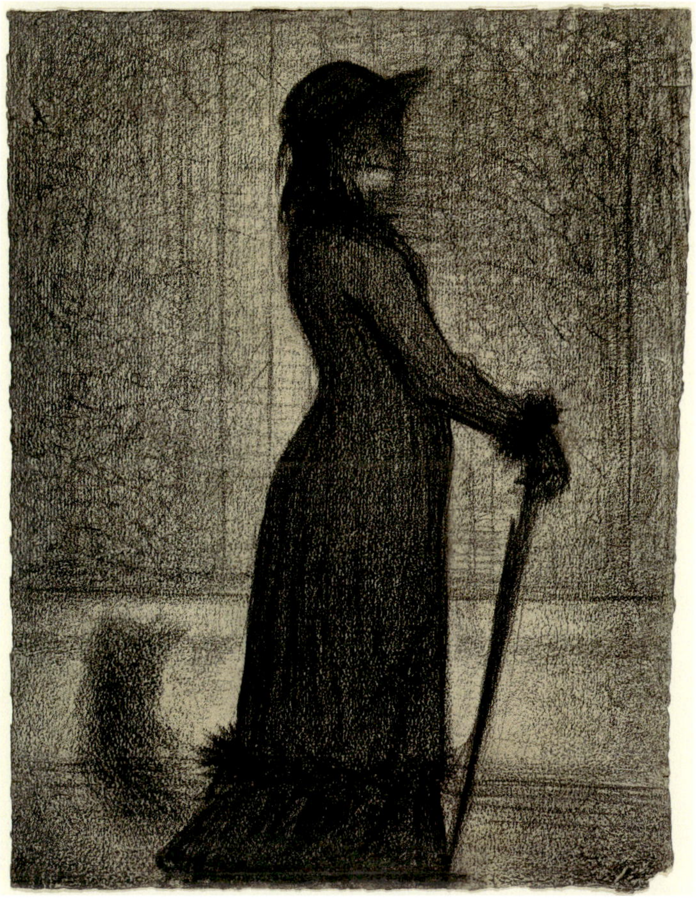

FIGURE 2.25.
Demonstration of hiding inpainting in high- and low-spatial-frequency regions of Cotes's *Portrait of Joseph Gulston and His Brother John Gulston*. (A) Four locations were selected, shown as red dots. (B) The painting's colors at these locations were increased in redness by the same amount while lightness was maintained. The color change is most noticeable in the sky, a region of low spatial frequency. As long as the lightness matches its surround, an inpainted area is nearly invisible in a region of high spatial frequency.

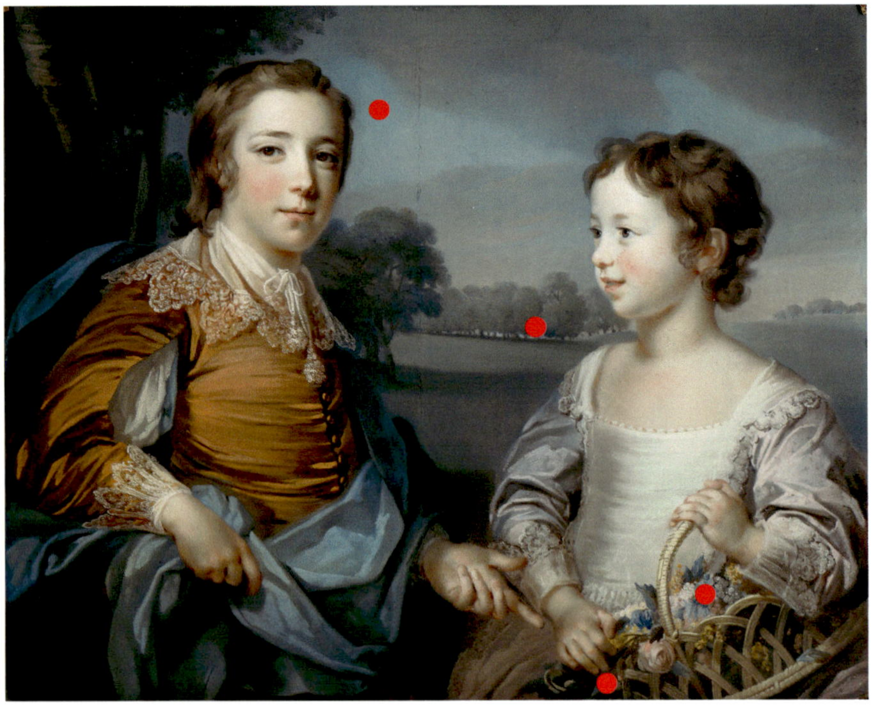

A

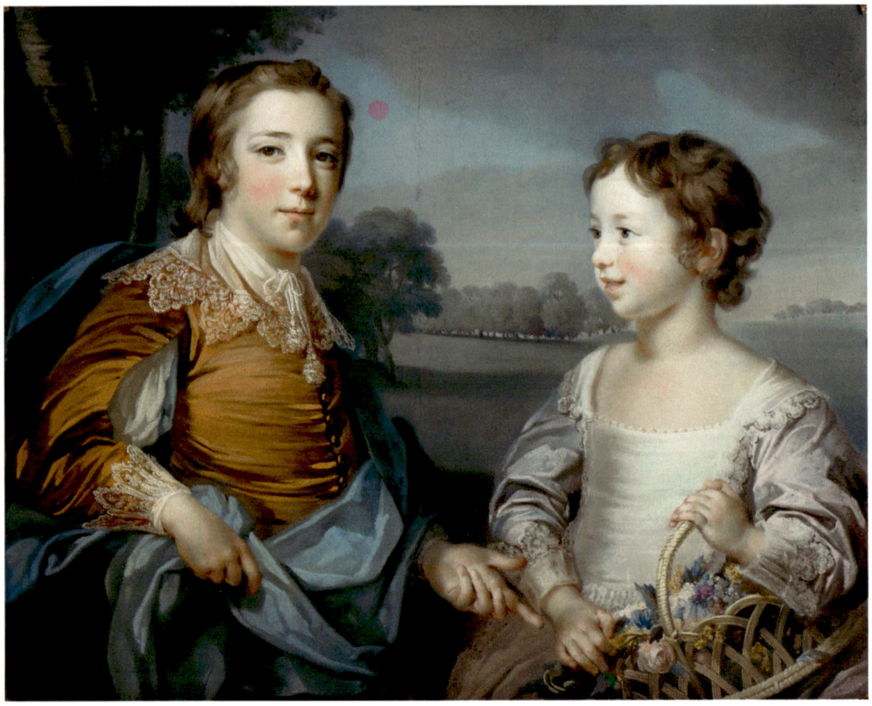

B

Summary

The human visual system is highly complex, processing light spectrally and spatially. At the first stage, our three cone receptors, L, M, and S, integrate incident light, reducing it from many wavelengths to three signals. Colors match when the three signals are equal; the spectral properties do not have to be identical. The three signals are further processed, in both the eye and the brain, through chromatic adaptation, signal compression, and color opponency. Some people have cone-receptor spectral sensitivities that are quite different from the norm or are completely missing, resulting in mild to severe color deficiencies. Differences in vision also occur as we age, leading to blurriness, changes in color vision, and reduced color discrimination. Opponent signals result from interconnecting cone receptors, which leads to visual illusions such as simultaneous contrast effects; this is particularly useful for extending an artist's palette and increasing a sense of depth. The interconnections result in high acuity for the black-opposed-to-white channel and low acuity for the red-opposed-to-green and yellow-opposed-to-blue channels. Artists often mimic our spatial vision in order to direct our gaze to specific passages. The success of imitative inpainting can depend on understanding our spatial vision.

Notes

1. More informally, "color-blind," though in fact only 0.005% of the population is completely color-blind; hence the more general term "color-deficient."
2. Unless listed otherwise, all the paint mixtures shown in this book used the following Golden Artist Color matte acrylic dispersion paints: rutile titanium dioxide white (PW 6), carbon black (PBk 7), cobalt blue (PB 28), ultramarine blue (PB 29), phthalocyanine blue (PB 15:4 green shade), phthalocyanine green (PG 7), permanent green light (PY 3 / PG 7), pyrrole orange (PO 73), arylide yellow (PY 74), pyrrole red (PR 254), dioxazine purple (PV 23), and quinacridone magenta (PR 122).
3. How can a signal be negative? The opponent signals are sent as pulses with respect to time. A neutral color will have a nominal pulse rate. A red stimulus will increase the rate, while a green stimulus will decrease it; thus, relative to the nominal rate, the signals are positive and negative.
4. The calculations for the simulation are in Brettel, Viénot, and Mollon 1997.
5. Figure 2.11 was adapted from Chevreul's plate II.
6. Figure 2.12 was inspired by the "Just GREY" example, © Gianni A. Sarcone, 1997–2013, www.giannisarcone.com, shown on the Archimedes Lab website: http://www.archimedes-lab.org/color_optical_illusions.html.

7. The illusion in figure 2.13, which was inspired by the home page of the International Colour Vision Society, occurs because the S cones have low acuity and may have a direct path to the brain. It is not an example of spatial mixing (assimilation), because the stripes and lettering are in focus and sharp (Monnier and Shevell 2003).
8. A low-pass spatial filter at 10 cycles per pixel was used for the low-frequency image. A band-pass spatial filter from 10 to 120 cycles per pixel was used for the medium-frequency image. A high-pass spatial filter at 120 cycles per pixel was used for the high-frequency image.
9. Mapping the achromatic data to a single gray value generated the chromatic image.
10. "Inpainting" and "restorative inpainting" are the terms used most frequently in the United States for this technique. In Europe, the term "retouching" is often used, although the historic practice of overpainting areas for aesthetic reasons, which is also called retouching, gives the term a negative connotation (Digney-Peer et al. 2012; Nadolny 2012). For the purposes of this book, two types of inpainting are considered. The first is imitative, where the goal is to have the inpainting match as closely as possible the appearance of the surrounding area. The second is discernible, where the goal is to achieve an appearance match at a distance but not close up.

References

Albers, J. 1963. *Interaction of Color*. New Haven, CT: Yale University Press.

Brettel, H., F. Viénot, and J. D. Mollon. 1997. "Computerized Simulation of Color Appearance for Dichromats." *Journal of the Optical Society of America* A14: 2647–55.

Chevreul, M. E. 1839. *De la loi du contraste simultané des couleurs et de l'assortiment des objets colorés*. First published in English 1854; republished 1967. *The Principles of Harmony and Contrast of Colors,* 16. New York: Reinhold.

Daw, N. 2012. *How Vision Works: The Physiological Mechanisms behind What We See*. New York: Oxford University Press.

DeValois, R. L., C. J. Smith, S. T. Kitai, and S. J. Karoly. 1958. "Responses of Single Cells in Different Layers of the Primate Lateral Geniculate Nucleus to Monochromatic Light." *Science* 127: 238–39.

Digney-Peer, S., K. Thomas, R. Perry, J. Townsend, and S. Gritt. 2012. "The Imitative Retouching of Easel Paintings." In *The Conservation of Easel Paintings,* edited by J. H. Stoner and R. Rushfield, 607–34. New York: Routledge.

Hering, E. 1878. *Zur Lehre vom Lichtsinne*. Republished in English 1964. *Outlines of a Theory of the Light Sense,* translated by L. M. Hurvich and D. Jameson. Cambridge, MA: Harvard University Press.

Hofer, H., J. Carroll, J. Neitz, M. Neitz, and D. R. Williams. 2005. "Organization of the Human Trichromatic Cone Mosaic." *Journal of Neuroscience* 25: 9669–79.

Hurvich, L. M., and D. Jameson. 1957. "An Opponent-Process Theory of Color Vision." *Psychological Review* 64: 384–404.

Judd, D. B. 1966. "Fundamental Studies of Color Vision from 1860 to 1960." *Proceedings of the National Academy of Sciences USA* 55: 1313–30.

Kevan, P. G., and W. G. K. Backhaus. 1998. "Color Vision: Ecology and Evolution in Making the Best of the Photopic Environment." In *Color Vision: Perspectives from Different Disciplines,* edited by W. G. K. Backhaus, R. Kliegl, and J. S. Werner, 163–83. Berlin: Walter de Gruyter.

Knoblauch, K., F. Saunders, M. Kusuda, R. Hynes, M. Podgor, K. E. Higgins, and F. M. de Monasterio. 1987. "Age and Illuminance Effects in the Farnsworth-Munsell 100-Hue Test." *Applied Optics* 26: 1441–48.

Kries, J. von. 1878. "Beitrag zur Physiologie der Gesichtsempfindungen." *Archiv für Physiologie* 2: 503–24. Republished in English 1970. "Contribution to the Physiology of Visual Sensations." In *Sources of Color Science,* edited by D. L. MacAdam, 101–8. Cambridge, MA: MIT Press.

Lindberg, D. C. 1976. *Theories of Vision from al-Kindi to Kepler*. Chicago: University of Chicago Press.

Marmor, M. F. 2006. "Ophthalmology and Art: Simulation of Monet's Cataracts and Degas' Retinal Disease." *Archives of Ophthalmology* 124: 1764–69.

Monnier, P., and S. K. Shevell. 2003. "Large Shifts in Color Appearance from Patterned Chromatic Backgrounds." *Nature Neuroscience* 6: 801–2.

Nadolny, J. 2012. "History of Visual Compensation for Paintings." In *The Conservation of Easel Paintings,* edited by J. H. Stoner and R. Rushfield, 573–85. New York: Routledge.

Sharpe, L. T., A. Stockman, H. Jagle, H. Knau, G. Klausen, A. Reitner, and J. Nathans. 1998. "Red, Green, and Red-Green Hybrid Pigments in the Human Retina: Correlations between Deduced Protein Sequences and Psychophysically Measured Spectral Sensitivities." *Journal of Neuroscience* 18: 10053–69.

Smith, V. C., and J. Pokorny. 1975. "Spectral Sensitivity of the Foveal Cone Photopigments between 400 nm and 500 nm." *Vision Research* 15: 161–71.

Spear, P. D. 1993. "Neural Bases of Visual Deficits during Aging." *Vision Research* 33: 2589–2609.

Stevens, S. S. 1961. "To Honor Fechner and Repeal His Law." *Science* 133: 80–86.

Svaetichin, G. 1956. "Spectral Responses from Single Cones." *Acta Physiologica Scandinavica* 39 (suppl. 134): 17–46.

Thoen, H. H., M. J. How, T.-H. Chiou, and J. Marshall. 2014. "A Different Form of Color Vision in Mantis Shrimp." *Science* 343: 411–13.

Werner, J. S. 1998. "Aging through the Eyes of Monet." In *Color Vision: Perspectives from Different Disciplines,* edited by W. G. K. Backhaus, R. Kliegl, and J. S. Werner, 3–41. Berlin: Walter de Gruyter.

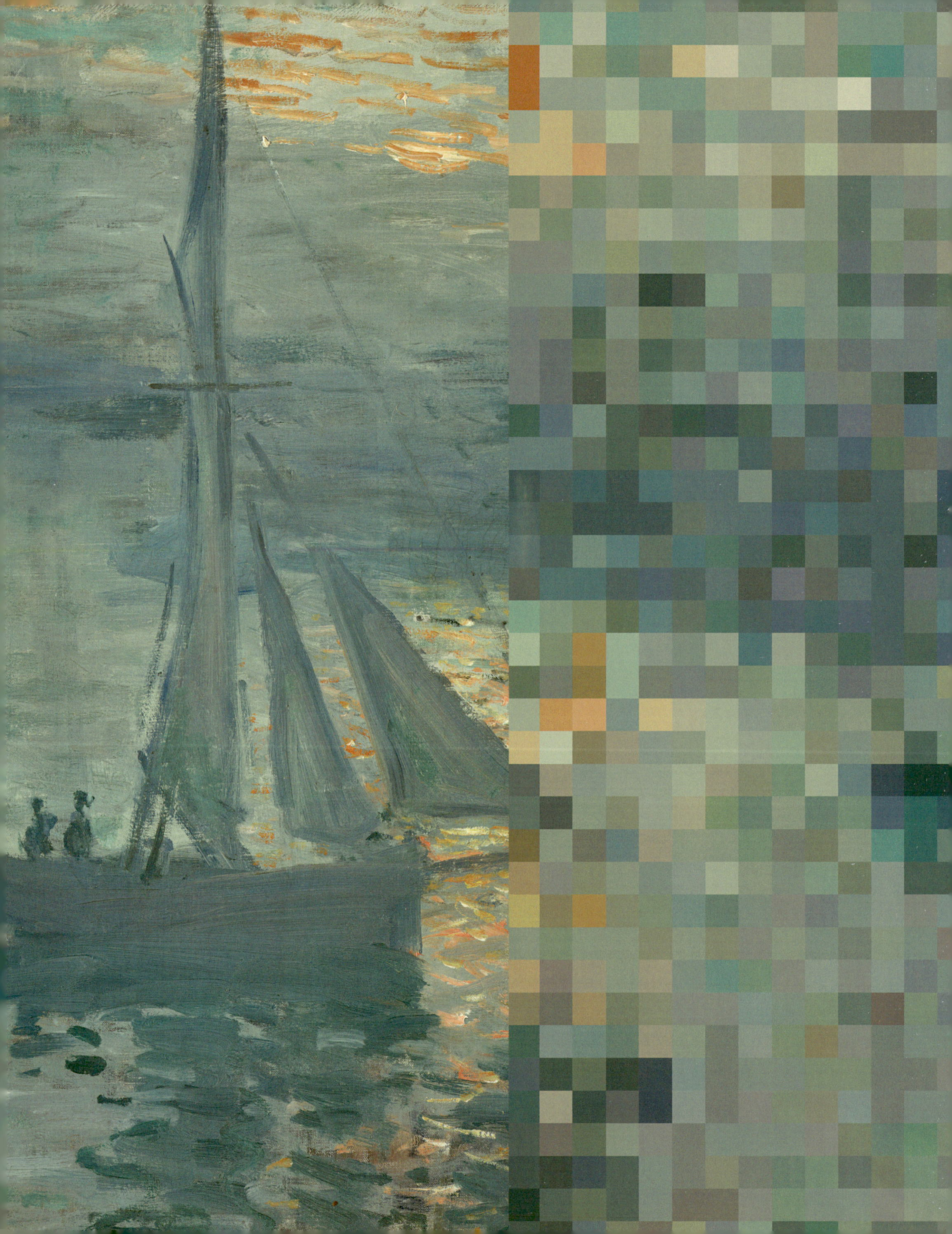

VISUAL AND NUMERICAL COLOR DESCRIPTIONS 3

This detail of Claude Monet's *Sunrise (Marine)* has been reduced in both its resolution and total number of unique colors in order to reveal the individual colors of the painting. Viewing this detail at an increasing distance demonstrates our visual acuity: eventually the pixelation is no longer seen (details, figs. 3.1a and b).

Describing colors and, more importantly, changes in color is difficult without a common point of reference. For centuries, artists have used pigments such as vermilion, ultramarine, and ocher as reference colors, but these terms lack specificity, as the range of colors so named by different paint manufacturers suggests. Two aspects of color, lightness and hue, are easy to describe and seem intuitive. We might speak of a dark green or a light blue, for instance, and if we say we have made a color redder, we understand that the visual effect depends on the color we started with. A redder red is more chromatic (closer to what we would think of as pure red), while a redder blue becomes purplish.

Producing samples with a set of visual attributes that vary systematically, with corresponding nomenclature, is one way to illustrate colors and color changes. With usage, a common visual description results and the nomenclature becomes a common language. But there are many situations in which greater color precision and accuracy are required, such as specifying color for manufacturing, quantifying color change that occurs over time, or evaluating the color quality of artwork reproduction. Precise description is achieved by numerical specification based on physical measurement.

Psychological Attributes

Imagine describing Claude Monet's *Sunrise (Marine),* shown in figure 3.1, to someone who hasn't seen the painting. "It is a painting of ships in a harbor at sunrise," you might say. "The sky is partly cloudy and the harbor is very gray. The sun, an intense orange, is peeking out through the clouds. It can also be seen reflected in the water." The two dominant colors in this description are gray and orange. It is easier to identify the colors if you eliminate the subject matter from consideration, which is accomplished by reducing the image resolution, and if you cut the total number of unique colors—in this case, to one hundred colors. Even with this simplification, any two of us might use very different words in our description, one person using specific pigment names like cadmium orange and viridian and the other relying on common color names like beige, orange, and green.[1] Neither approach is exact.

Colors seen in a single viewing situation can be categorized by three psychological attributes: hue, lightness, and chromatic intensity (Kuehni and Schwarz 2008). Note that the grays in *Sunrise* range in hue from neutral to greenish; the colors vary in lightness, some quite dark and some very pale; and the oranges show a range of chromatic intensity from beige to fiery orange. In figure 3.2 the one hundred colors are arranged by each of the three attributes in turn, with the attribute changing systematically.[2] None of these scales seems well ordered, however, since the other attributes also vary. A visual scale should vary along a single dimension while the others remain constant. Within each pair of adjacent samples there should be a visual difference that is similar to that of the next pair. But visualizing more than one attribute would require arranging colors in two or three dimensions, combining, for instance, hue and chromatic intensity, or hue, lightness, and chromatic intensity.

Hue is a visual attribute that enables colors to be categorized as red, yellow, green, or blue or as a combination of two adjacent colors when they are arranged in a circle in this order (CIE 2011). Because of our opponent color vision, a color will not look simultaneously red and green or simultaneously yellow and blue. Thus colors can readily be categorized as a combination of red and yellow, yellow and green, green and blue, or blue and red. The Natural Colour System uses this categorization, defining a color's hue as a percentage of adjacent hues, for example, 90% yellow and 10% red (Hård, Sivik, and Tonnquist 1996a, b). Defining hue in this manner

FIGURE 3.1.
(A) Claude Monet (French, 1840–1926), *Sunrise (Marine),* 1873. Oil on canvas, 48.9 × 58.4 cm (19¼ × 23½ in.). Los Angeles, J. Paul Getty Museum, 98.PA164. (B) Reducing the image resolution and the total number of unique colors to one hundred reveals the individual colors of the painting.

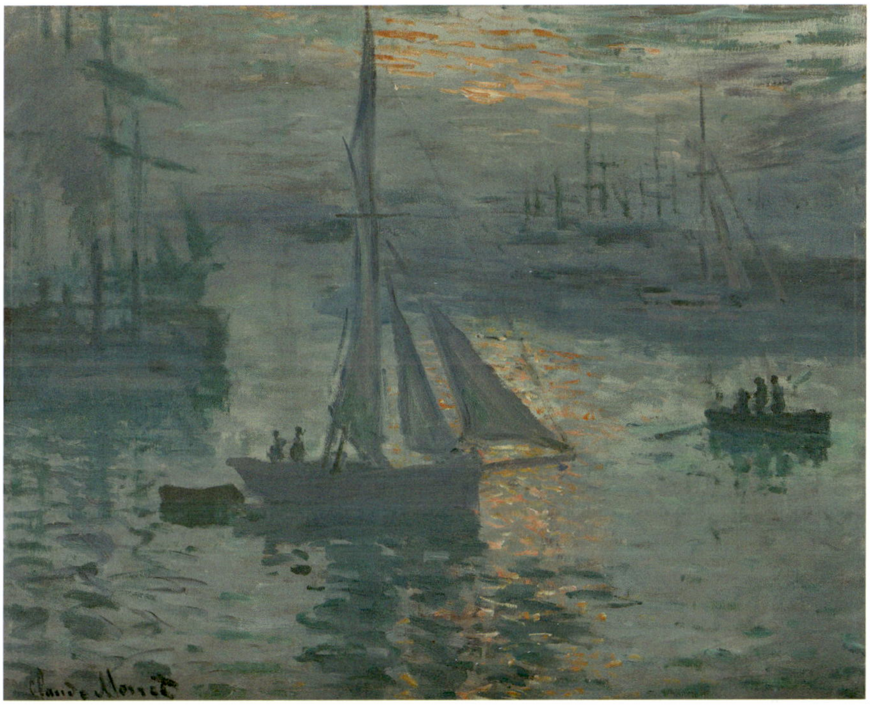

A

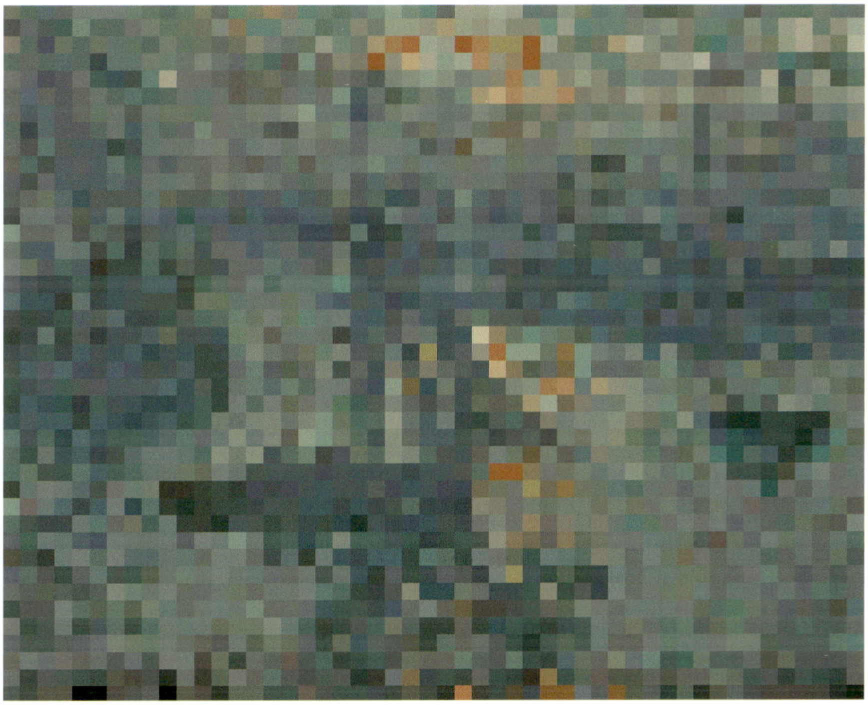

B

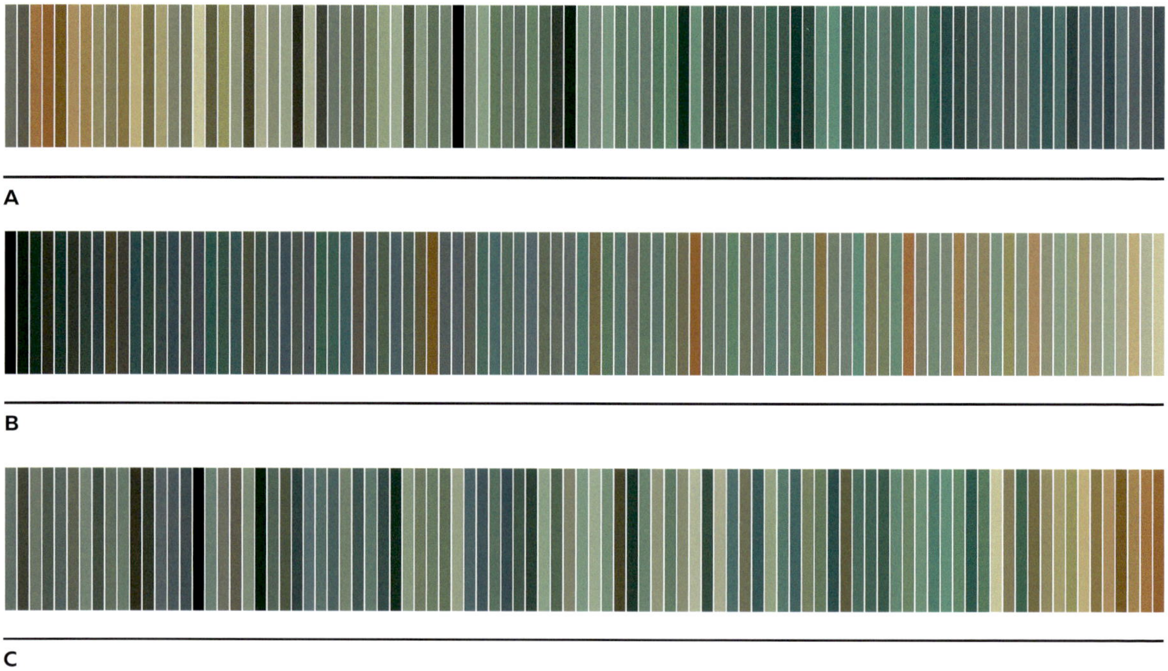

FIGURE 3.2. The one hundred colors in Monet's *Sunrise (Marine),* ordered by (A) hue, (B) lightness, and (C) chromatic intensity.

relies on our ability to visualize fundamental hues, known as unique hues, devoid of adjacent hues (Mollon and Jordan 1997). These can be thought of as psychological primaries rather than mixing primaries, for example, red, yellow, and blue; or cyan, magenta, yellow, and black; or red, green, and blue. The concept of a unique hue is illustrated in figure 3.3 with twelve yellows that range from reddish to greenish. There should be one yellow that appears neither reddish nor greenish.[3]

We have an intuitive understanding of hue, and we often arrange colors of various hues naturally in a circle. It is more problematic to decide how many hues to include and how to place them relative to one another around the circle. The difficulty arises from our tendency to use color wheels for color mixing, not for defining perception. Varying the hue at three different levels of constant lightness and chromatic intensity is shown in figure 3.4.

Lightness, the second attribute of color, indicates a color's similarity to a neutral color ranging from black to white (ASTM 2012), shown in figure 3.5 for three colors—bluish red, gray, and green—at increasing

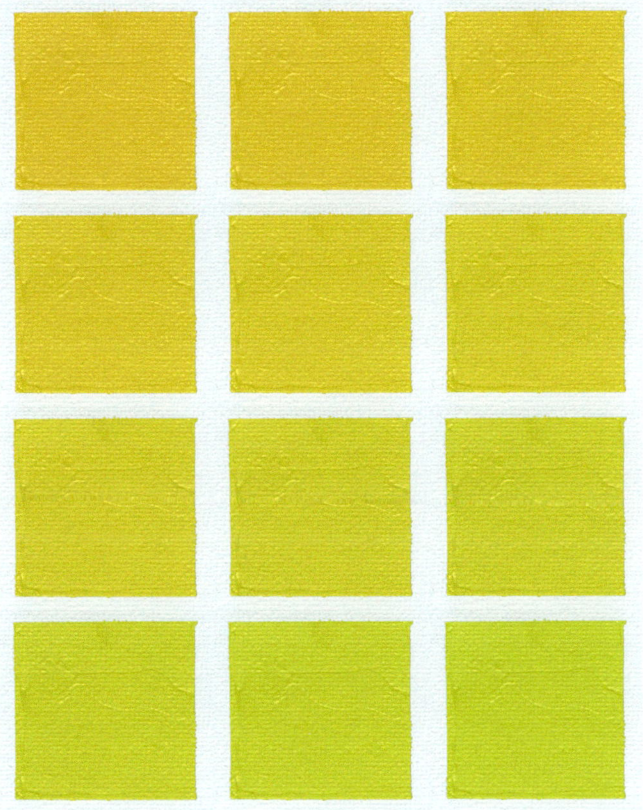

FIGURE 3.3. Yellows ranging from reddish to greenish. One of these squares—but not the same one for all viewers—should appear to be unique yellow, that is, a yellow that has no redness or greenness in it.

levels of lightness. Envisioning a color becoming lighter or darker is certainly intuitive.

The third attribute of color is chromatic intensity, defined as either the amount of color or the degree of the departure from neutral gray. Chromatic intensity is shown in figure 3.6 for an orange color when mixed with white, gray, or black. Moving from left to right, chromatic intensity increases. The mixtures with white and black also vary in lightness, while the mixture with gray has a constant lightness.

There are several ways to arrange hue, lightness, and chromatic intensity in three dimensions: as a cone with black at its point, as a double cone with black at one point and white at the other, and as a vertical cylinder with black at one end and white at the other (Kuehni and Schwarz

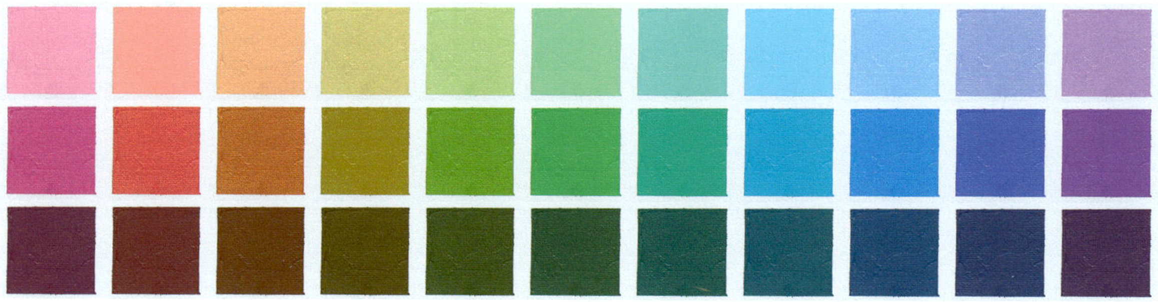

FIGURE 3.4. Eleven hues at three different levels of constant lightness and chromatic intensity.

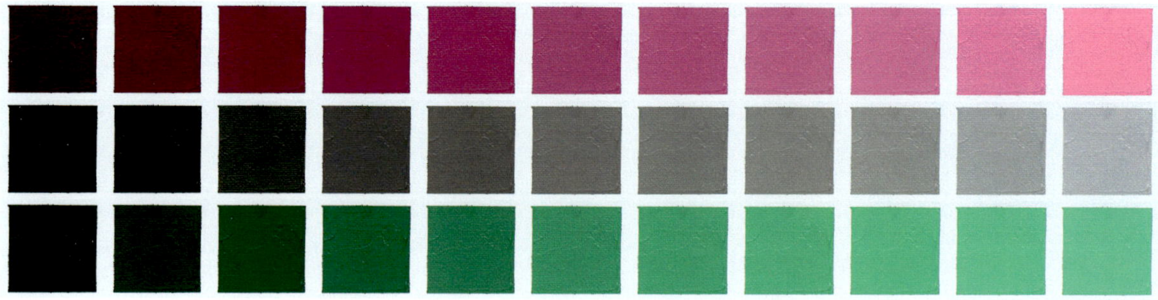

FIGURE 3.5. Changes in lightness (increasing from left to right) for three colors—bluish red (top row), neutral (middle), and green (bottom).

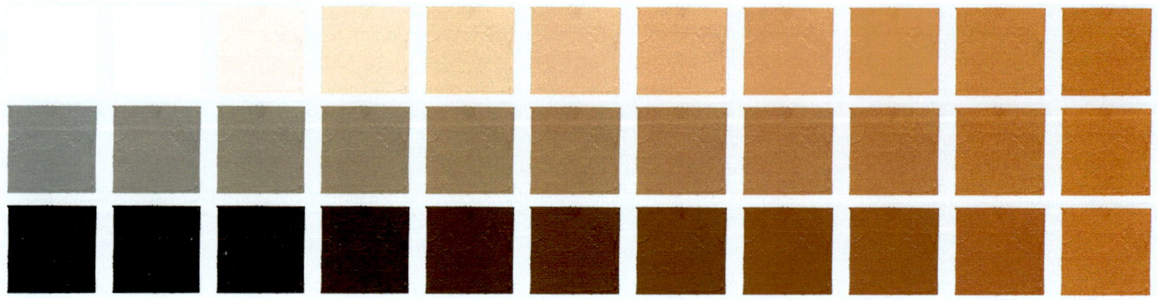

FIGURE 3.6. Changes in chromatic intensity (increasing from left to right) for an orange color mixed with white (top row), gray (middle), or black (bottom).

2008). Hue revolves around each neutral gray central axis radially. A color order system that is based on a double cone is the Natural Colour System (Hård, Sivik, and Tonnquist 1996a, 1996b). Its origins trace back to Hering and opponent color-vision theory. For a specific hue, colors are arranged in a triangular grid with blackness, whiteness, and chromaticness at the three apexes, as shown in figure 3.7. Colors are assigned as a percentage of blackness, whiteness, and chromaticness.

In 1899 the artist, educator, and inventor Albert H. Munsell developed a cylindrical system (Kuehni 2002; Landa and Fairchild 2005). For a specific hue, colors are arranged in a rectangular grid like the one shown in figure 3.8, with value (lightness) as its vertical axis and chroma as its horizontal axis. A color that changes in Munsell chroma does not vary in lightness or hue. In essence, chroma is chromatic intensity at a constant lightness. Thus chroma is an attribute of color used to indicate the degree of departure of the color from a neutral color of the same lightness (ASTM 2012).

Both the Natural Colour System and the Munsell system are used for color specification, the latter, for example, by Golden Artist Colors. Each has a physical atlas of colors arranged by hue where adjacent samples have approximately equal visual spacing. This facilitates visual interpolation and gaining familiarity with visual color specification.

Defining a color by its lightness and chroma takes practice because colors often vary in both attributes simultaneously, as in the case of a three-dimensional colored object lit from a single direction. Psychological terms can be denoted as a "ness" (Engeldrum 2000) — for example, lightness and darkness. These terms are often opponent: redness or greenness, yellowness or blueness. As shown in figure 3.9, three "nesses" can be used to define colors within a plane of lightness and chromatic intensity: paleness or depth(ness), grayness or chroma(ness), and dullness or vividness (Berns 2014). These terms can be used to define either a color or a change in color. Making a color grayer requires reducing chromatic intensity without changing lightness. Making a color duller requires reducing both lightness and chromatic intensity. A vivid color has both high lightness and high chromatic intensity. These attributes combine features of both the Natural Colour System and the Munsell system. Comparisons of the three scales are shown in figure 3.10.[4]

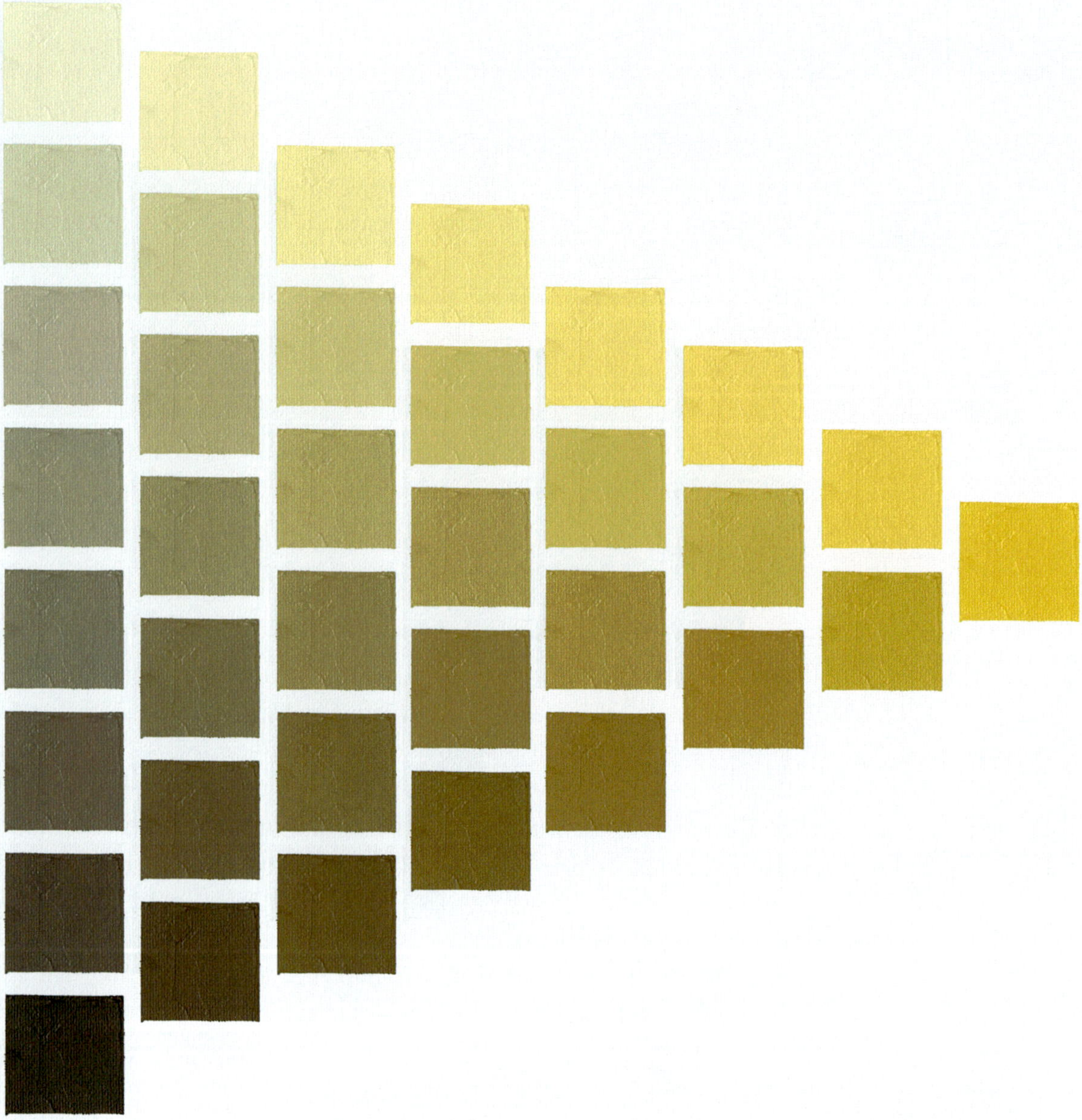

FIGURE 3.7. Yellows arranged in a triangular grid.

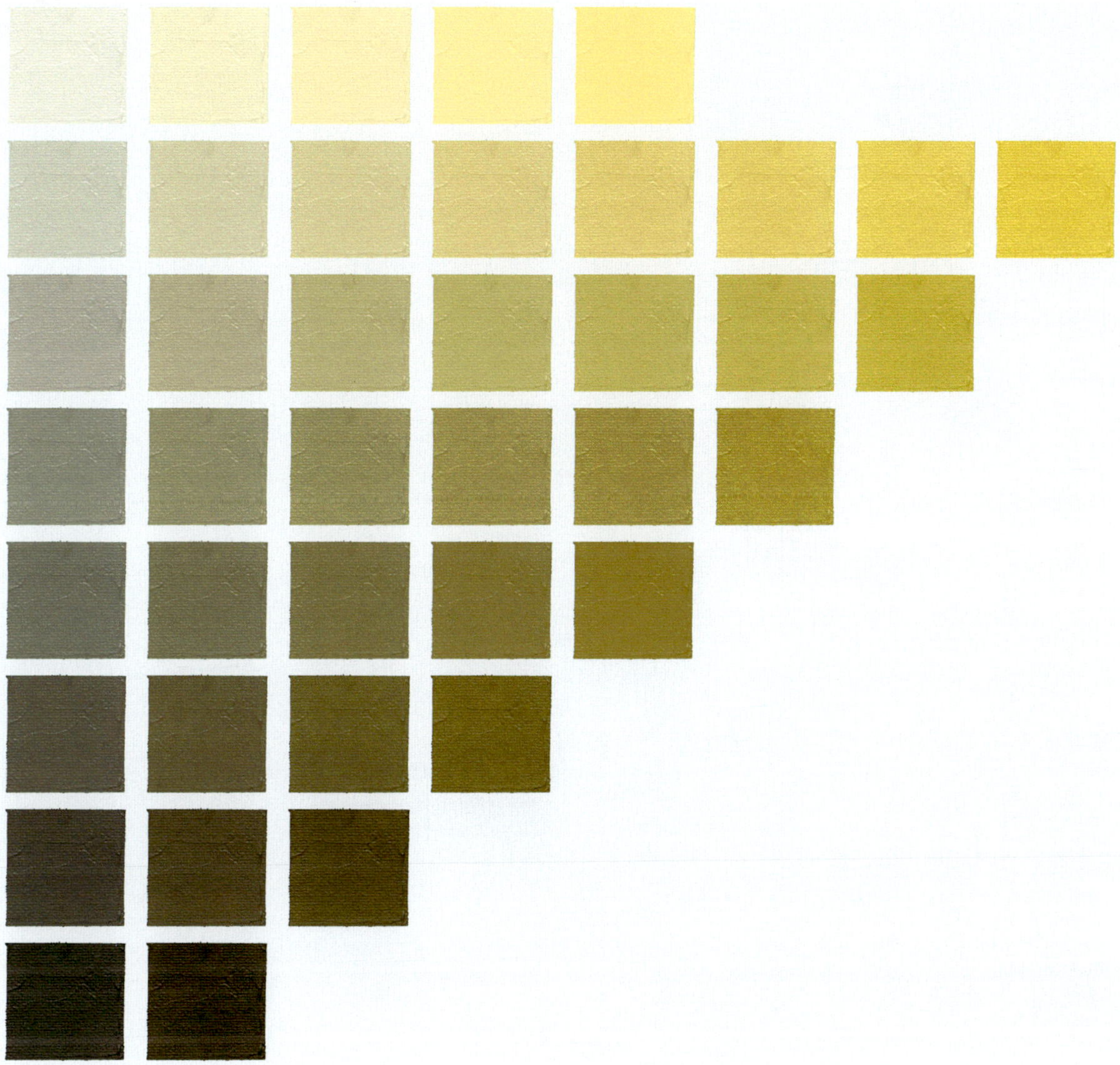

FIGURE 3.8. Yellows arranged in a rectangular grid. The maximum achievable chroma for each lightness row depends on the particular yellow colorant and our visual system.

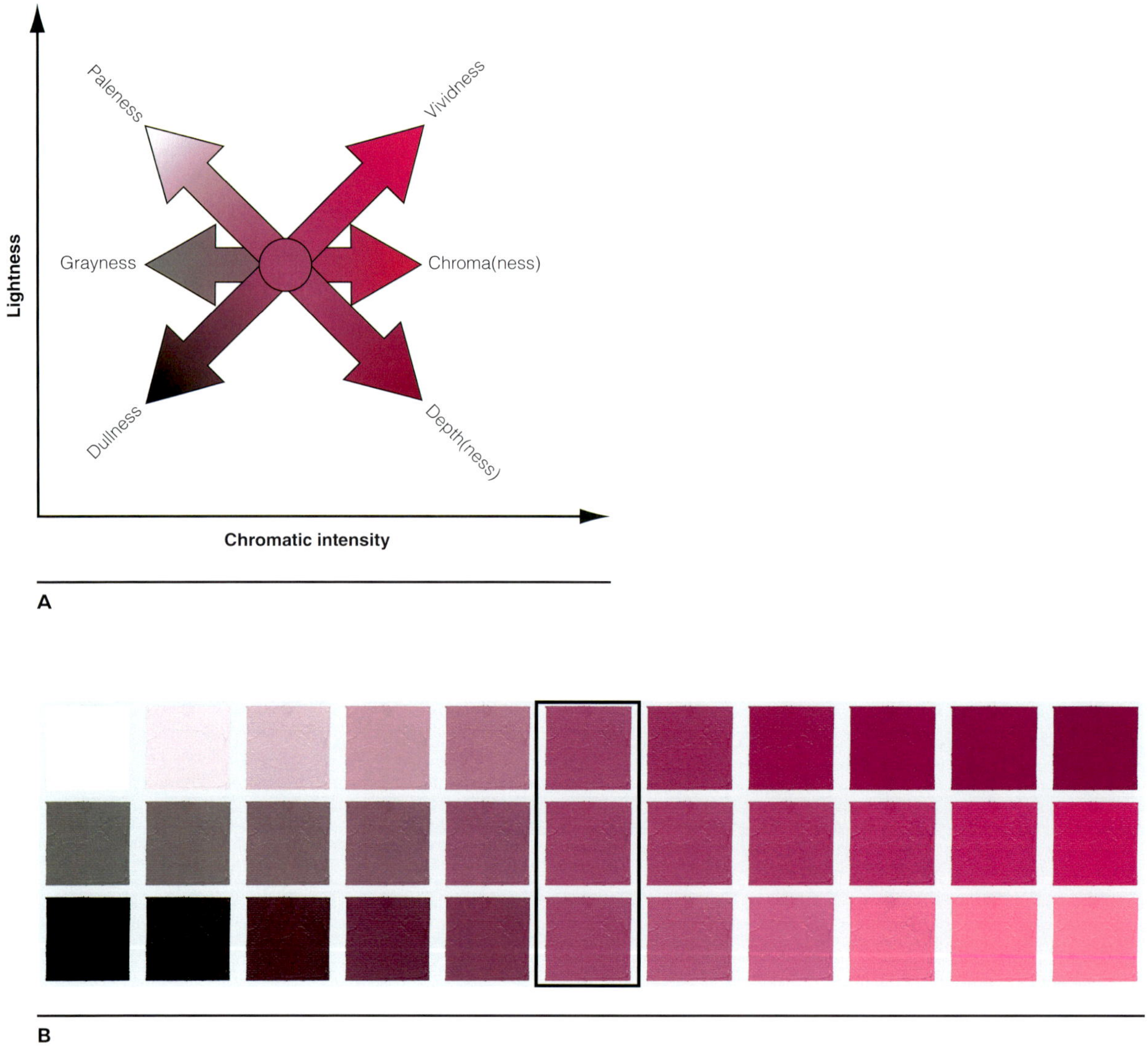

FIGURE 3.9. (A) The "nesses" of colors at a constant hue. (B) Colors were produced in which each attribute—paleness or depth(ness) in the top row, grayness or chroma(ness) in the middle row, and dullness or vividness in the bottom row—varied about a central color (outlined in black).

FIGURE 3.10. Scales of (A) depth(ness), (B) chroma(ness), and (C) vividness for various hues (increasing from left to right).

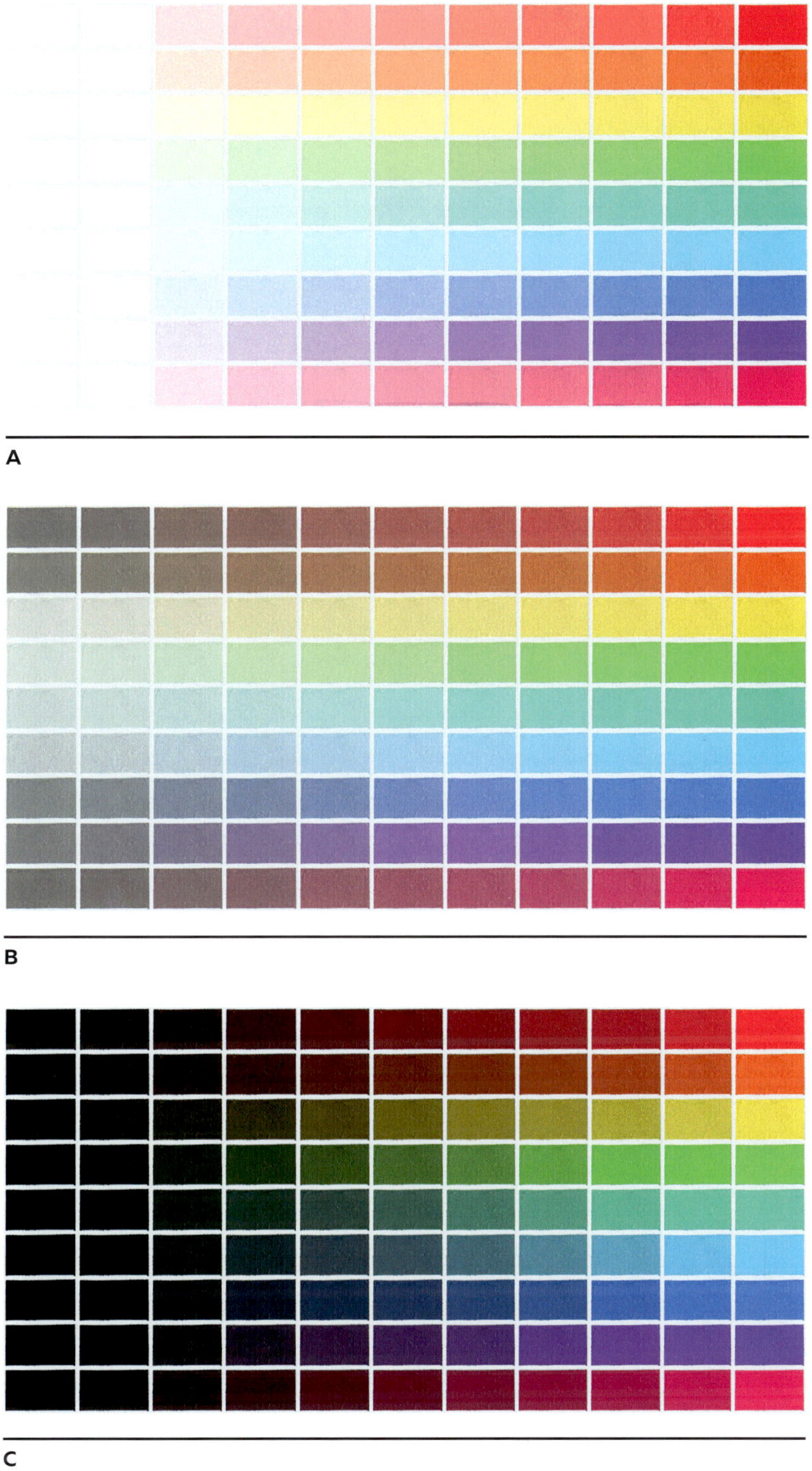

CIE L, a*, and b* (CIELAB)*

CIELAB (pronounced *see-lab*) is a numerical system that is used internationally for color description (CIE 2004). CIE is the acronym for Commission Internationale de l'Eclairage (International Commission on Illumination), a group that sets the standards and recommended practices for lighting, vision, color, and imaging. The colors of most manufactured materials are defined using CIELAB. Color reproduction uses "Lab" alongside RGB and CMYK as color spaces. CIELAB is used extensively in conservation science.

Color vision is composed of four stages: integration of incident light into trichromatic signals for our L, M, and S cone receptors; chromatic adaptation; nonlinear signal compression; and color opponency. CIELAB, too, incorporates these four stages. When CIE developed the CIELAB system during the 1970s, however, there were no standardized L, M, and S spectral sensitivities. Instead, there were standardized "color-matching functions" based on metameric color matching.

Suppose we use three primary colors to make a range of color mixtures. Artists might choose red, yellow, and blue along with white paint, while physicists would choose red, green, and blue lights. Mixing lights has two advantages for visual experiments: a larger range of colors and a mixing system that can change colors continuously. The physics approach was used for experiments performed in the 1920s, in which observers were shown two adjacent fields of colored light. In one field, varying the intensities of red, green, and blue lights produced color. The other field was a single wavelength mixed with white light, the white added to reduce chromatic intensity. The task was to make a color match between the two areas by adjusting the three-color mixture. The process was repeated for each wavelength throughout the visible spectrum and for seventeen observers. Standardizing three lights and averaging the observers' matches resulted in a standard set of color-matching functions, known as the CIE 1931 standard observer.[5] For reasons of practical importance in 1931, the three lights were theoretical, leading to the use of X, Y, and Z rather than RGB. The three functions are denoted in figure 3.11 as $\bar{x}$, $\bar{y}$, and $\bar{z}$, where the bar indicates an average. Comparing X, Y, and Z color-matching functions with the L, M, and S spectral sensitivities reinforces the theoretical nature of this system: each of the visual-system spectral sensitivities has a single peak while the $\bar{x}$ color-matching function has two peaks, and the shapes of $\bar{y}$ and $\bar{z}$ are different from those of M and S. A color is defined by its

FIGURE 3.11. Comparison of (A) color-matching functions of the CIE 1931 standard observer, $\bar{x}$ (dashed), $\bar{y}$ (solid), and $\bar{z}$ (dot-dashed), with (B) spectral sensitivities of the human visual system's L (orange), M (green), and S (blue) cone receptors.

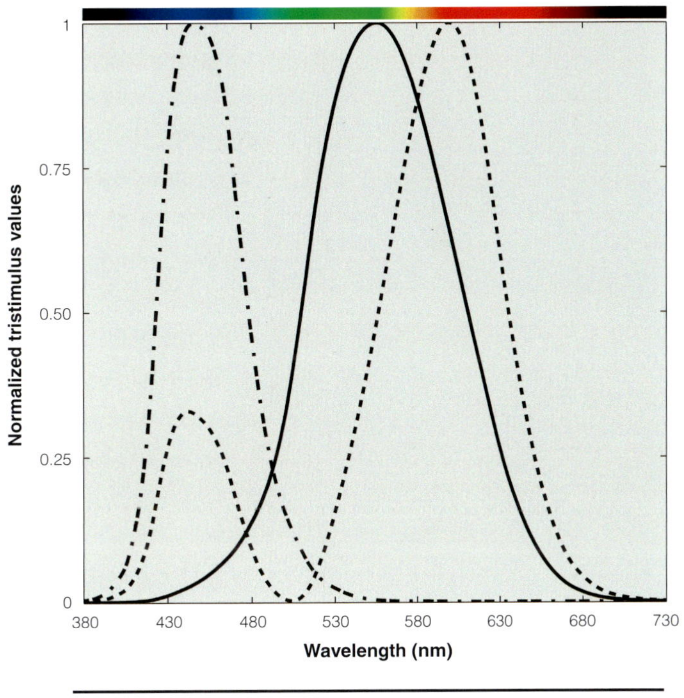

A

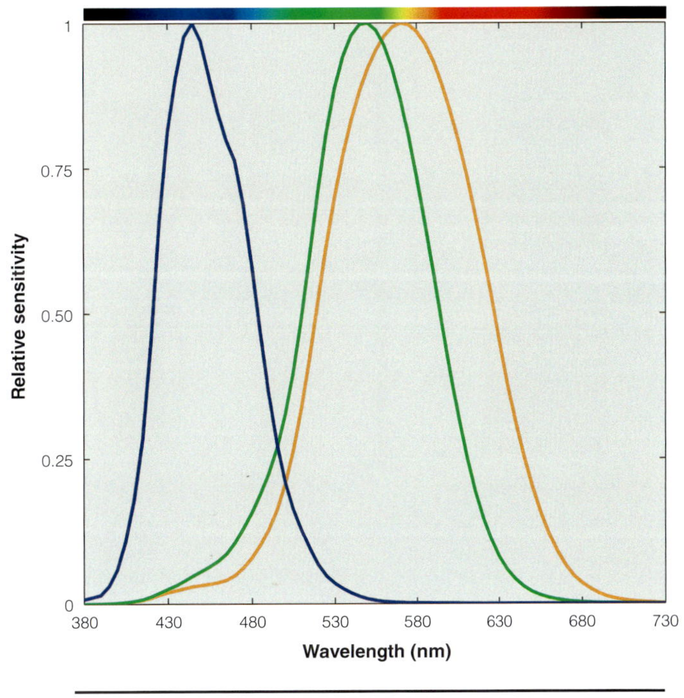

B

tristimulus values, X, Y, and Z, or simply XYZ. The XYZ system is also a trichromatic system, based on integrating incident light in much the same way that our cone receptors do. Spectrophotometers built for color specification report XYZ tristimulus values with the operator deciding on the lighting under which the object will be evaluated — for example, indoors lit by incandescent light or outdoors in direct sun — and choosing a specific observer, usually a CIE standard observer. These choices depend on the application.

To illustrate these points, a target was made, sampling hue and lightness, and imaged spectrally. The 36-band data were used to calculate an XYZ image of the target when illuminated by sunlight and viewed by the 1931 standard observer. An XYZ image was also calculated for a 100% reflecting white under identical illumination. The two tristimulus images were used to calculate an L*a*b* image, with the mathematics shown in figure 3.12 and the image in figure 3.13. The black-and-white (achromatic)

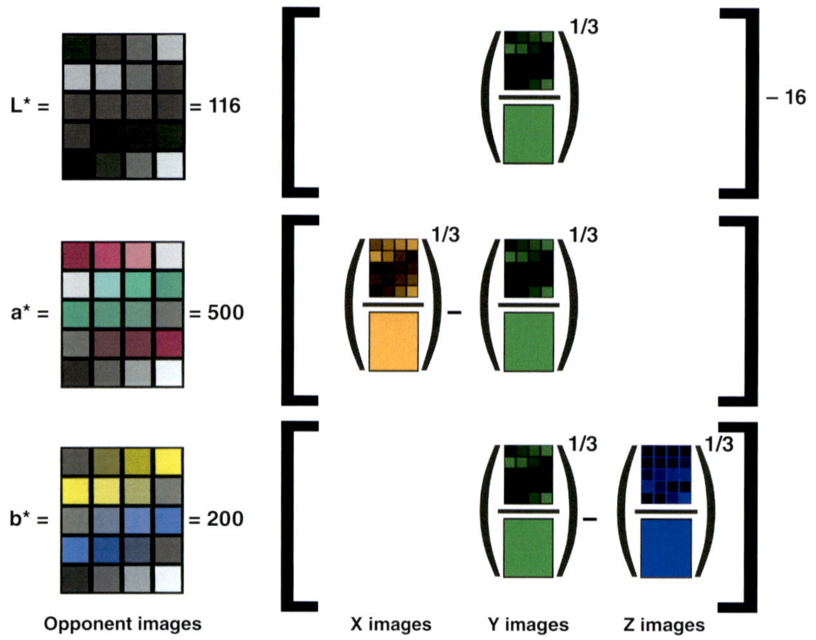

FIGURE 3.12. Mathematics of calculating CIE L*, a*, and b* from XYZ tristimulus values. Chromatic adaptation is accomplished by dividing the object's tristimulus values by the tristimulus values of a 100% reflecting white. A cube root is used for compression. The subtractions produce opponent signals. The constants of 116 and 16 scale L* from 0 (black) to 100 (white). The constants of 500 and 200 scale the opponent signals toward a visually uniform color space.

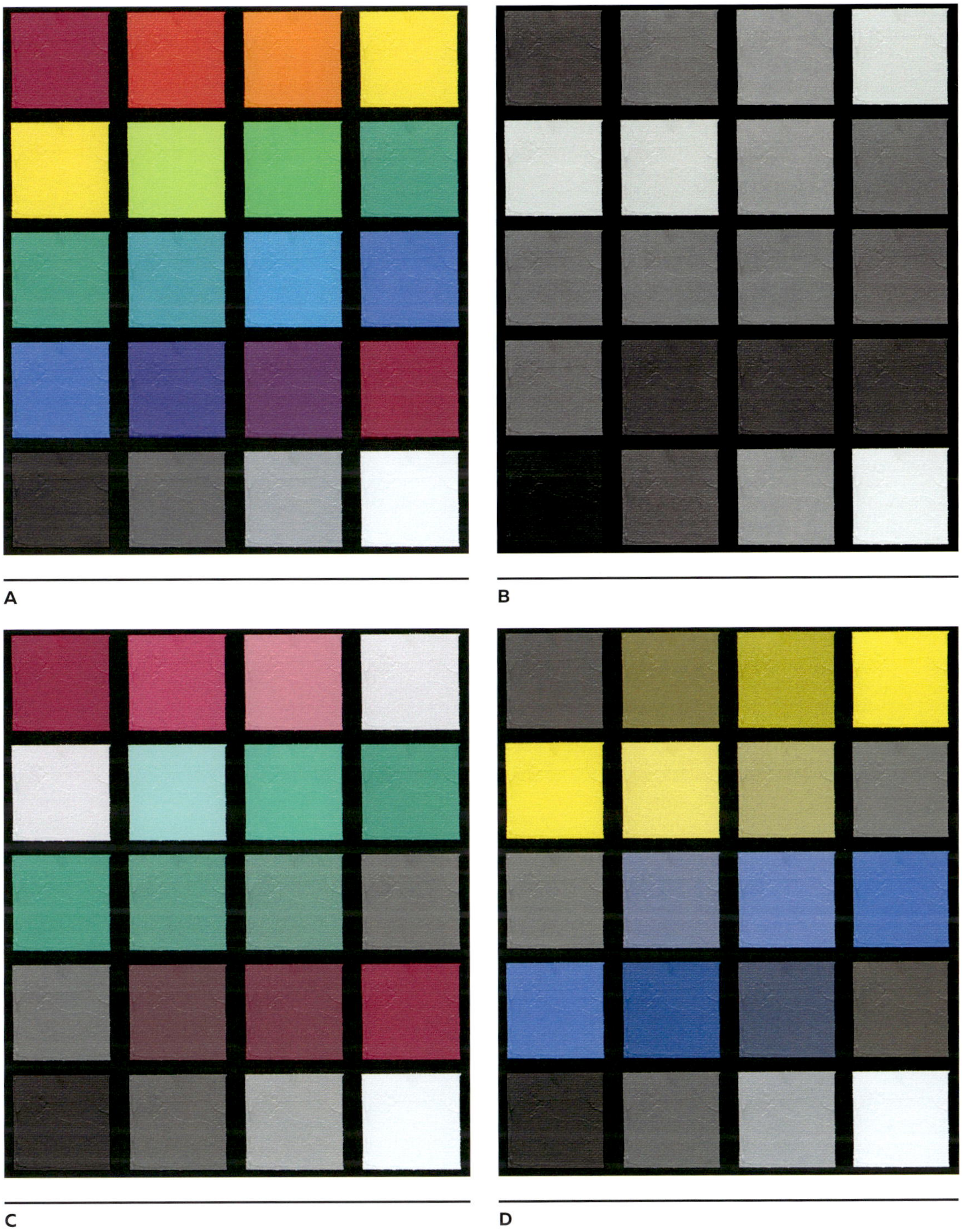

FIGURE 3.13. (A) Hue and lightness target, with (B) L*, (C) a*, and (D) b* images.

channel is a lightness signal, L* (pronounced *el-star*). The a* and b* signals are chromatic: the a* signal crudely corresponds to redness or greenness, where +a* is redness and –a* is greenness, and the b* signal corresponds reasonably well to yellowness or blueness, where +b* is yellowness and –b* is blueness. The adjective "crudely" is used in connection with the a* image because the colors look closer to magenta and bluish green than to the expected red and green. The difference can be seen in figure 3.14, where the visual system's red-opposed-to-green channel is compared with the a* channel for Henri Rousseau's *A Centennial of Independence*.

L*a*b* coordinates define a color numerically. It is possible to visualize colors solely from their coordinates, particularly for near neutrals. As colors increase in chromatic intensity, however, visualization becomes more difficult, as we are mentally plotting the a*b* coordinates to determine a color's hue. When CIELAB was under development, color scientists well versed in the Munsell system recognized that transforming the rectangular a*b* coordinates to cylindrical-polar coordinates would lead to descriptions similar to Munsell hue and chroma (Robertson 1990). The radial distance from the neutral axis defines CIELAB chroma, C^*_{ab}, while the angle, measured counterclockwise from the positive a* axis, defines CIELAB hue, h_{ab}. The rectangular and cylindrical-polar coordinates of the hue and lightness target are listed in figure 3.15 and graphed in CIELAB looking down the lightness axis.[6] The target's hues were sampled at 30° increments. Colors with hue angles of 30° and 150° have hues similar to unique red and green.

Returning to Monet's *Sunrise (Marine)*, in figure 3.16 the one hundred colors are plotted in CIELAB. The three-dimensional graph produces a "color gamut." By looking down the L* axis, we can plot the data as a two-dimensional graph with a* and b* axes. A limitation of looking down the L* axis is that dark colors can be obscured by lighter colors, as a comparison of the two- and three-dimensional graphs shows. This is remedied by plotting L* and C^*_{ab}. Thus, a color can be defined using L*, a*, and b* or L*, C^*_{ab}, and h_{ab}.

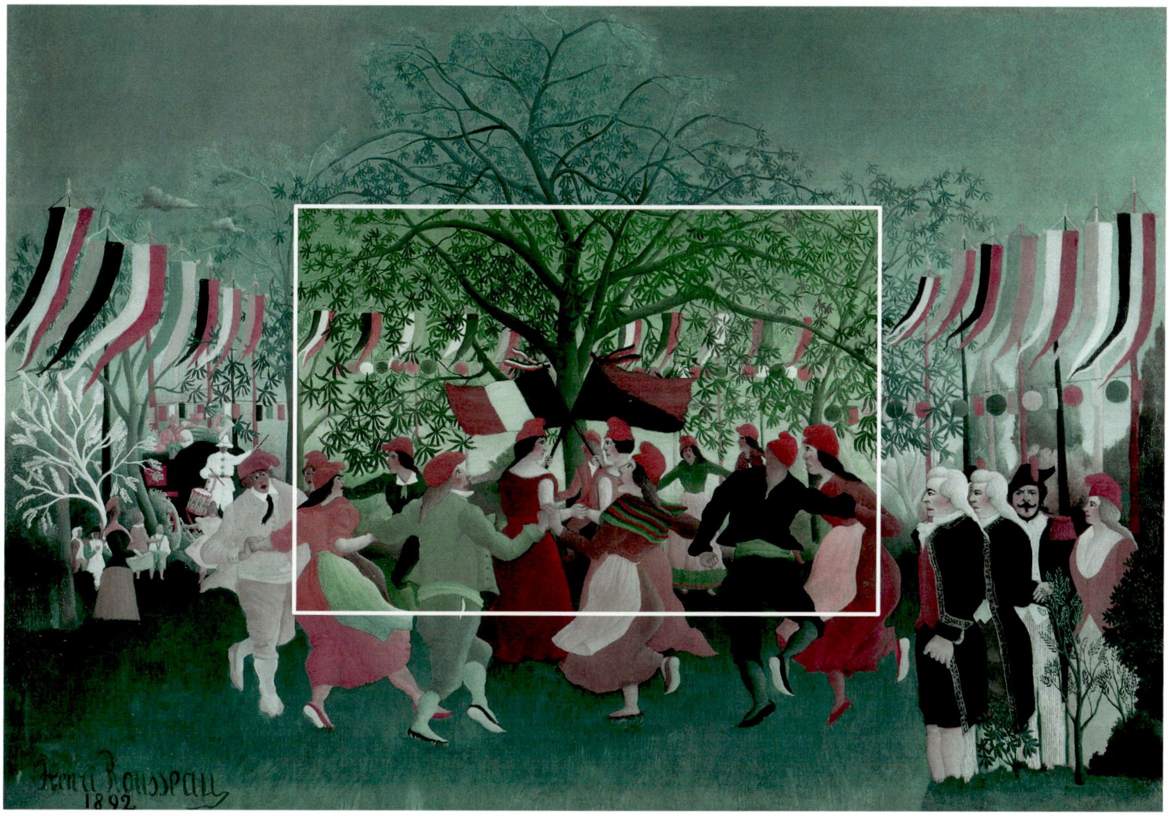

FIGURE 3.14. Henri Rousseau (French, 1844–1910), *A Centennial of Independence,* 1892. Oil on canvas, 111.8 × 157.2 cm (44 × 61⅞ in.). Los Angeles, J. Paul Getty Museum, 88.PA58. The painting is divided into areas rendered as our visual system's red-opposed-to-green channel (center) and CIELAB's a* (outside). CIELAB a* is not equivalent to redness or greenness; rather, it is magenta(ness) or bluish greenness.

FIGURE 3.15. (A) Nominal CIELAB coordinates of the color target, and (B) color target plotted in CIELAB looking down the lightness axis. (The vivid hue circle is shown as an aid to relate a* and b* coordinates with hue. Colors can plot beyond this circle.)

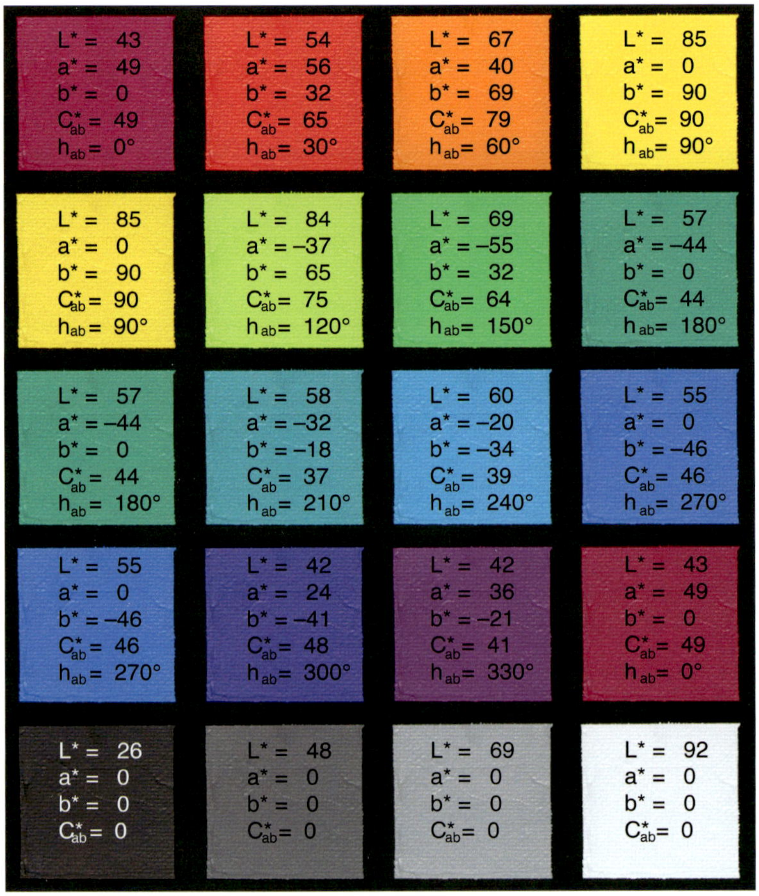

A

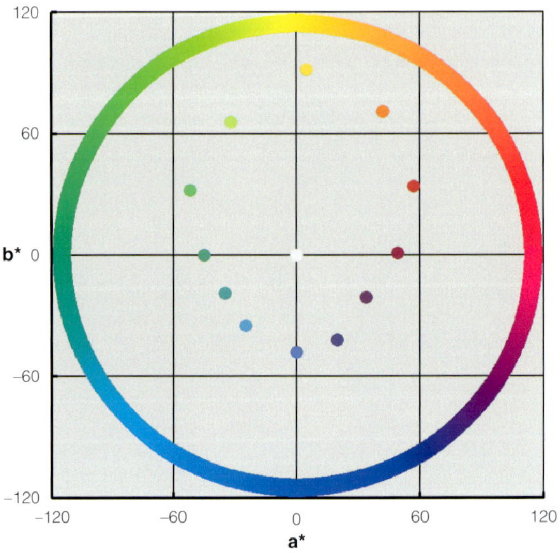

B

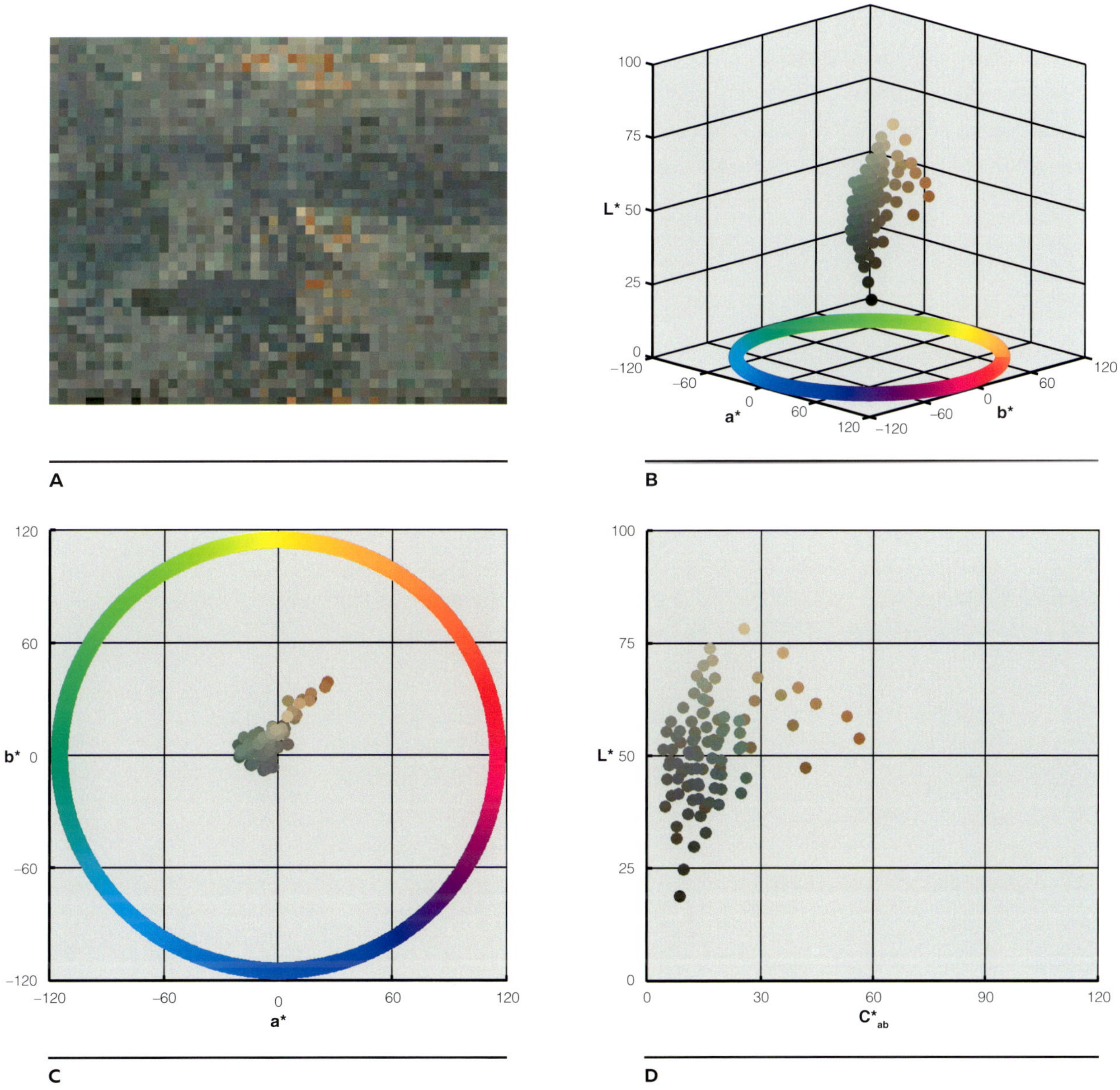

FIGURE 3.16. (A) The one hundred colors of Claude Monet's *Sunrise (Marine)* plotted in CIELAB (B) as a three-dimensional plot, (C) looking down the lightness axis, and (D) as a lightness and chroma plot.

Color Differences and Tolerances

All manufacturing specifications include tolerances, and color is no exception. When we purchase a tube of paint, we assume that its color, both out of the tube and in mixture, is identical to the color of the last tube we bought from the same manufacturer, whether yesterday, last month, or last year. CIELAB can be used to define the paint's color, where each batch of paint must match the specification within a tolerance. Since CIELAB has three coordinates, tolerances can be specified for each coordinate—for example, $\Delta L^* = \pm 2$, $\Delta a^* = \pm 2$, and $\Delta b^* = \pm 2$, where Δ refers to difference and the reference color is subtracted from the test color. Any difference greater than 2 or smaller than −2 is out of tolerance. As noted above, a* and b* and, in turn, Δa^* and Δb^* are most useful for near-neutral colors. Tolerances can also be specified for chroma and hue, ΔC^*_{ab} and ΔH^*_{ab}.[7]

A tolerance—expressed as either ΔL^*, Δa^*, and Δb^* or ΔL^*, ΔC^*_{ab}, and ΔH^*_{ab}—forms a box about a color specification. However, the distances from the center of the box to its surface are not constant, making such a tolerance problematic. Instead of a box, a sphere is used: the distance from the sphere's center to its surface is the same. The radius of the sphere defines the tolerance. In color technology, the term "total color difference" is used to represent this radius. To generalize, the three-dimensional distance between any two colors is their total color difference. There is a symbol to represent this distance: ΔE^*_{ab} (pronounced *delta-e-star-a-b*). We would say that the tolerance for a color specification is a particular total color difference. As a distance, ΔE^*_{ab} defines only the magnitude of the color difference, not the direction. Sometimes, however, the direction of change is more important than its magnitude. In this case it is important to evaluate either ΔL^*, Δa^*, and Δb^* or ΔL^*, ΔC^*_{ab}, and ΔH^*_{ab} in addition to ΔE^*_{ab} and, if possible, plot the two colors.

As manufacturers of colored products gained experience using ΔE^*_{ab} for color tolerances, a systematic lack of correlation with visual assessment emerged where high-chroma colors could have larger ΔC^*_{ab} and ΔH^*_{ab} tolerances than low-chroma colors. In addition, as these tolerances increased, the ΔC^*_{ab} tolerance was three times larger than the ΔH^*_{ab} tolerance. The net result was that increases in the standard's chroma changed the tolerance from spherical to ellipsoidal. In 1994 the CIE recommended a new formula for color tolerances that accounted for this chroma dependency, CIE94 (CIE 1995). Additional systematic trends were found

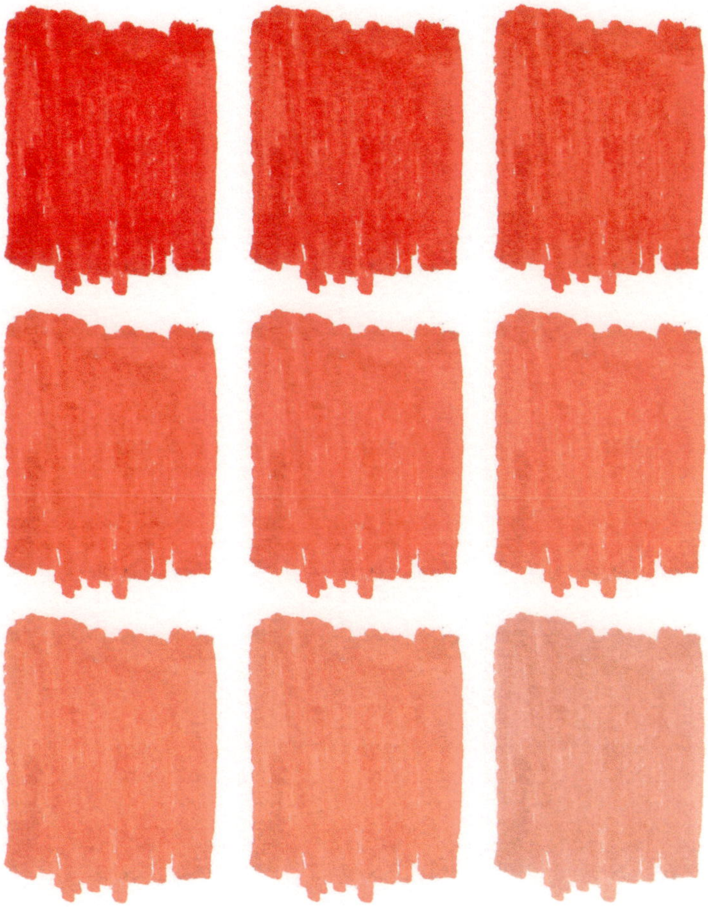

FIGURE 3.17. Simulating the effects caused by excessive light exposure for a marker. The total color difference ranges between 5 ΔE^*_{ab} and 40 ΔE^*_{ab} in 5 ΔE^*_{ab} increments when compared to the marker color before light exposure (top left).

that depended on lightness and hue, and in 2000 the CIE updated its 1994 recommendation, notated as either CIEDE2000 or ΔE_{00} (CIE 2004). Choosing between ΔE^*_{ab} and ΔE_{00} depends on existing standards and application.

Color differences are used in art conservation science to quantify how an object's color can change over time from light exposure, heat, and humidity (Whitmore, Pan, and Bailie 1999; Morris and Whitmore 2007). In figure 3.17 the color fading of a marker drawn on watercolor paper was simulated in 5 ΔE^*_{ab} increments, resulting in a range from 5 to 40 ΔE^*_{ab} compared with the unfaded marker.

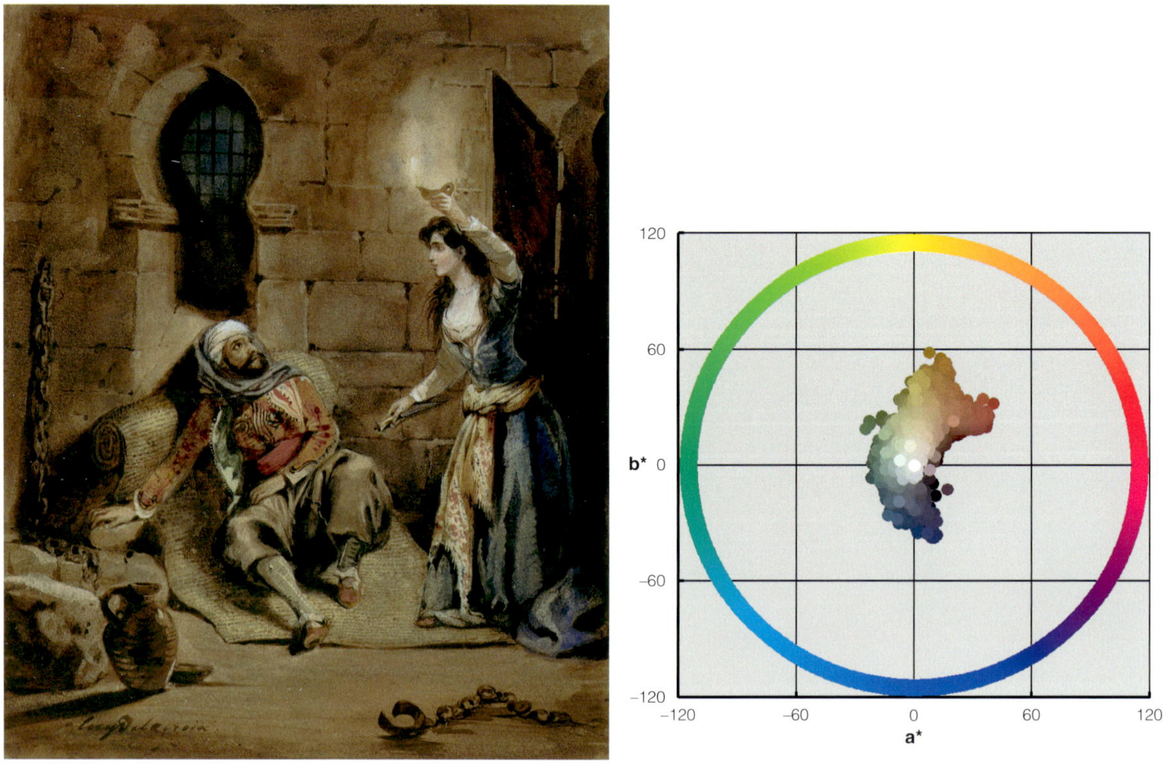

A

FIGURE 3.18. Eugène Delacroix (French, 1798–1863), *Episode from "The Corsair" by Lord Byron,* ca. 1831. Watercolor and brown ink with touches of gouache and a graphite underdrawing, 24.3 × 19.2 cm (9⅝ × 7½ in.). Los Angeles, J. Paul Getty Museum, 97.GC.30. Either (A) the man's red sash or (B) the entire watercolor was reduced in depth by 15 ΔE^*_{ab} to simulate fading. Next to each image, the CIELAB data are plotted looking down the lightness axis. Although physically identical, the sash appears deeper in color when the entire watercolor is reduced in depth than when only the sash is reduced in depth.

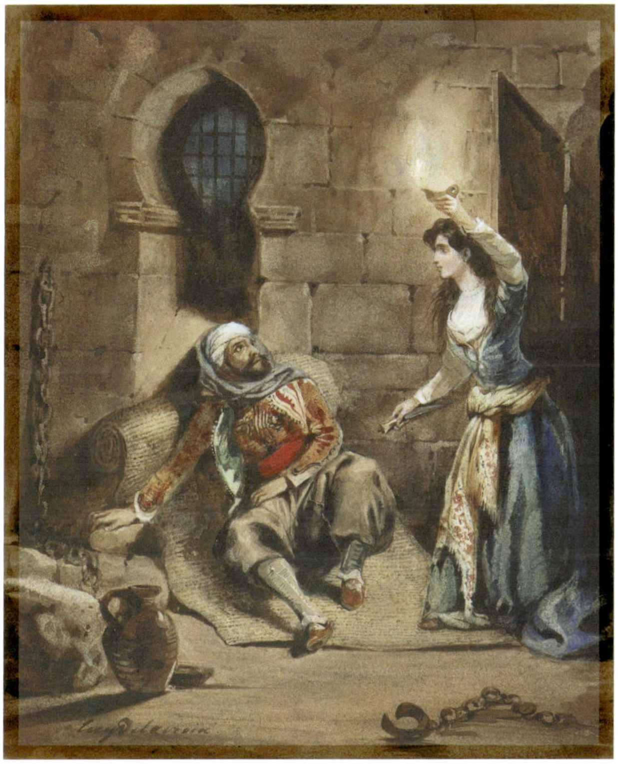

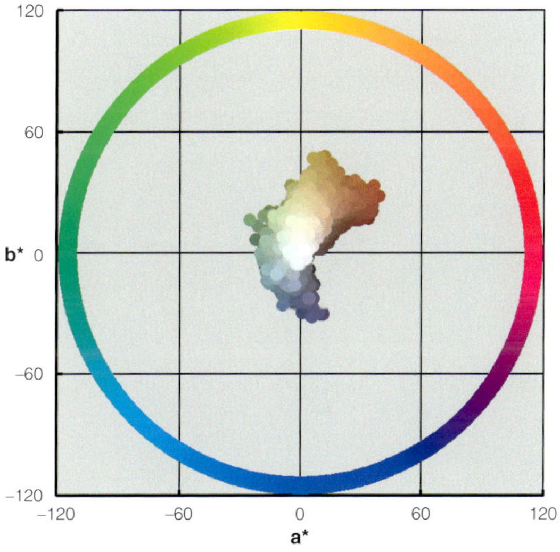

B

Determining an acceptable amount of fading depends on particular aspects of the composition—for example, whether it is figurative or abstract, whether the fading is color specific or over the entire artwork, and what its value is (Richardson and Saunders 2007). In figure 3.18 such changes were simulated for Eugène Delacroix's *Episode from "The Corsair" by Lord Byron,* where depth was reduced by 15 ΔE^*_{ab} for either the entire watercolor or the red sash only. Although fading the entire watercolor results in a large reduction in color gamut, as seen in the CIELAB plots, the red sash appears to have a greater chromatic intensity compared with its appearance when selectively faded.

Summary

Colors can be described using colorant names, common names, psychological attributes, color order systems, and numerics. The effectiveness of any particular approach depends on those who are communicating, their familiarity with the set of terms, and the precision and accuracy required. The attributes of redness, greenness, yellowness, blueness, lightness, chroma(ness), grayness, depth(ness), paleness, vividness, and dullness can be used to describe a color, while redder, greener, yellower, bluer, lighter, darker, more chroma, grayer, deeper, paler, more vivid, and duller describe changes in color. For numerical color specification, CIELAB is used, either the rectangular coordinates of L^*, a^*, and b^* or the cylindrical-polar coordinates of L^*, C^*_{ab}, and h_{ab}. When defining tolerances or the difference between two colors, ΔL^*, Δa^*, Δb^*, ΔC^*_{ab}, ΔH^*_{ab}, ΔE^*_{ab}, and ΔE_{00} are used.

Notes

1. Color naming is a fascinating subject, particularly as it relates to anthropology (Kay and Regier 2006) and frequency of usage (Moroney 2003).
2. These colors are ordered by CIELAB L^*, C^*_{ab}, and h_{ab}, representing lightness, chromatic intensity, and hue, respectively.
3. A range of yellows is shown because of variability among color-normal observers (Kuehni 2012).
4. Scales of depth(ness) and vividness derive from rectangular axes of lightness and chroma. Figures 3.9 and 3.10 were calculated using CIELAB's L^* (lightness) and C^*_{ab} (chroma) (Berns 2014).
5. The 1931 standard observer was based on visual judgments restricted to the center of the eye, specifically, a 2° field of view. For many industries, judgments are based on a larger field of view, and new experiments were performed during the 1950s resulting in the 1964 standard observer, subtending a 4° through 10° field of view.
6. The listed coordinates are based on image data. Measurements of the printed book will vary.
7. ΔH^*_{ab} is a distance between two colors that vary in hue, whereas Δh_{ab} is a difference in hue angle between two colors. For a fixed hue-angle difference, ΔH^*_{ab} increases with increases in chroma, C^*_{ab}. A color is specified by its L^*, C^*_{ab}, and h_{ab}. Its tolerance is specified by ΔL^*, ΔC^*_{ab}, and ΔH^*_{ab}. ("H^*_{ab}" has no meaning.)

References

ASTM. 2012. *Standard Terminology of Appearance (E284-12)*. West Conshohocken, PA: ASTM International.

Berns, R. S. 2014. "Extending CIELAB: Vividness, V^*_{ab}, Depth, D^*_{ab}, and Clarity, T^*_{ab}." *Color Research and Application* 39: 322–30.

CIE. 1995. *Industrial Colour Difference Evaluation (116-1995)*. CIE Technical Report. Vienna: Commission Internationale de l'Eclairage.

——. 2004. *Colorimetry (15:2004)*. 3rd ed. CIE Technical Report. Vienna: Commission Internationale de l'Eclairage.

——. 2011. *International Lighting Vocabulary (S 017/E:2011 ILV)*. Vienna: Commission Internationale de l'Eclairage.

Engeldrum, P. G. 2000. *Psychometric Scaling: A Toolkit for Imaging Systems Development*. Winchester, MA: Imcotek.

Hård, A., L. Sivik, and G. Tonnquist. 1996a. "NCS, Natural Color System – From Concept to Research and Applications. Part I." *Color Research and Application* 21: 180–205.

——. 1996b. "NCS, Natural Color System – From Concept to Research and Applications. Part II." *Color Research and Application* 21: 206–20.

Kay, P., and T. Regier. 2006. "Language, Thought and Color: Recent Developments." *Trends in Cognitive Sciences* 10: 51–54.

Kuehni, R. G. 2002. "The Early Development of the Munsell System." *Color Research and Application* 27: 20–27.

——. 2012. "Unique Hues and Their Stimuli – State of the Art." *Color Research and Application* 39: 279–87.

Kuehni, R. G., and A. Schwarz. 2008. *Color Ordered: A Survey of Color Order Systems from Antiquity to the Present*. Oxford: Oxford University Press.

Landa, E. R., and M. D. Fairchild. 2005. "Charting Color from the Eye of the Beholder." *American Scientist* 93: 436–43.

Mollon, J. D., and G. Jordan. 1997. "On the Nature of Unique Hues." In *John Dalton's Colour Vision Legacy*, edited by C. Dickinson, I. Murray and D. Carden, 381–92. London: Taylor & Francis.

Moroney, N. 2003. "Unconstrained Web-Based Color Naming Experiment." In *Color Imaging VIII: Processing, Hardcopy, and Applications; Proceedings of SPIE,* edited by E. Reiner and G. G. Marcu, 36–46. Bellingham, WA: SPIE.

Morris, H. R., and P. M. Whitmore. 2007. "'Virtual Fading' of Art Objects: Simulating the Future Fading of Artifacts by Visualizing Micro-fading Test Result." *Journal of the American Institute for Conservation* 46: 215–28.

Richardson, C., and D. Saunders. 2007. "Acceptable Light Damage: A Preliminary Investigation." *Studies in Conservation* 52:177–187.

Robertson, A. R. 1990. "Historical Development of CIE Recommended Color Difference Equations." *Color Research and Application* 15: 167–70.

Whitmore, P. M., X. Pan, and C. Bailie. 1999. "Predicting the Fading of Objects: Identification of Fugitive Colorants through Direct Nondestructive Lightfastness Measurements." *Journal of the American Institute for Conservation* 38: 395–409.

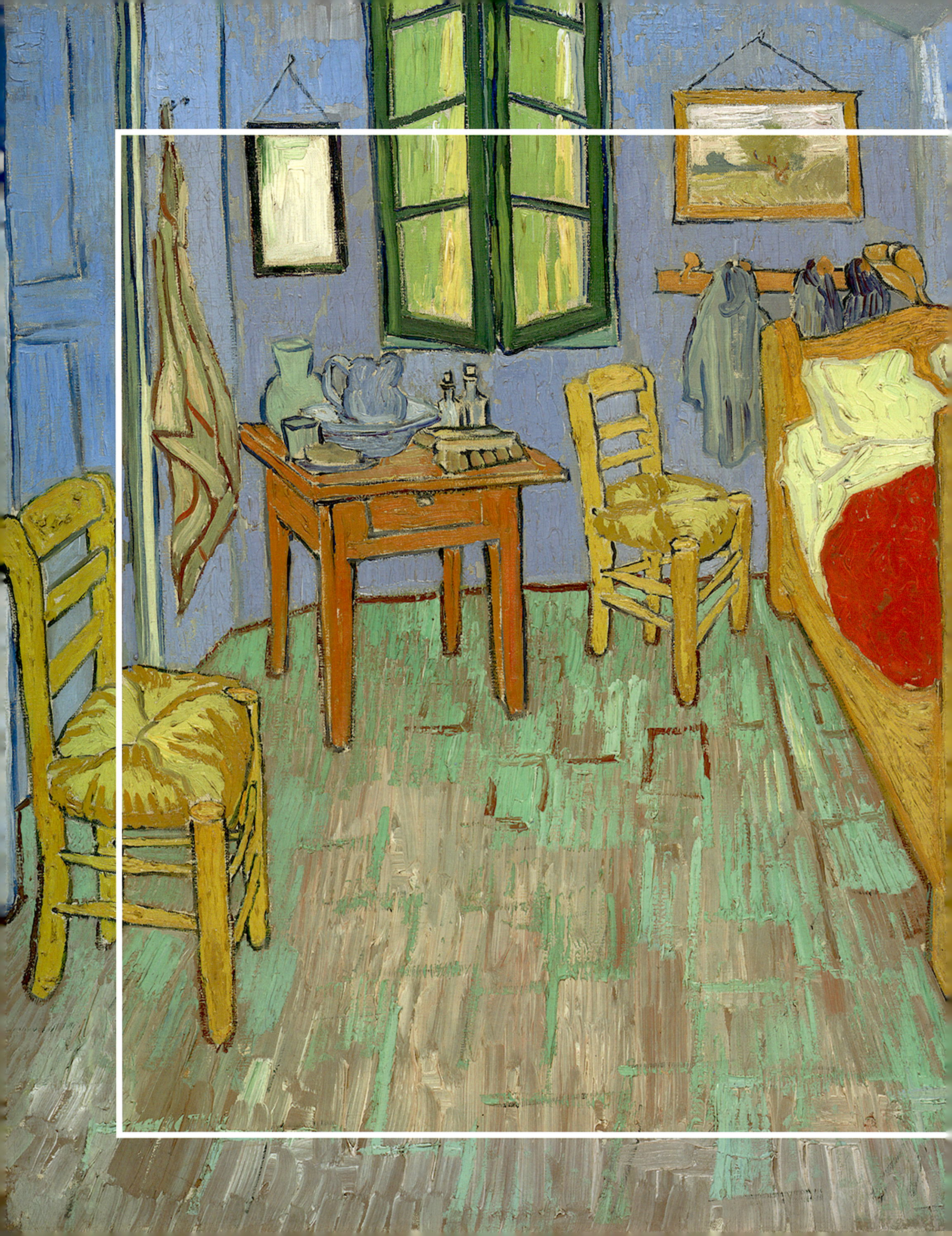

METAMERISM AND COLOR INCONSTANCY 4

Vincent van Gogh's *Bedroom* illuminated by either daylight (outer frame) or incandescent light (center). The artist's use of cobalt blue results in color inconstancy (detail, fig. 4.8a)

There are many situations in which we encounter two materials that match in color but are clearly composed of different colorants. The paintings reproduced in this book have been printed with cyan, magenta, yellow, and black inks. It is highly unlikely, however, that any of the artists used this palette. As a consequence, the spectral properties of the printed reproductions are very different from those of the paintings. Such spectral variations can cause changes in color-matching quality as lighting is changed; this is known as metamerism. Imitative inpainting can become discernible under different lighting when colorants are chosen that do not closely match the spectral properties of the surrounding area. Another undesirable effect of altered lighting is color inconstancy, in which an object's color changes as lighting varies. This can occur, for example, when an artist's studio has natural daylight and a gallery uses incandescent lighting. Color relationships carefully crafted in the studio may not persist when the artwork is exhibited.

Metamerism

The word "metamerism" comes from the chemical term "metamer." One of the most important concepts in color science, metamerism is the phenomenon in which spectrally different stimuli match in color. Because of the trichromatic nature of color vision, colors match when the identical LMS cone signals are produced, even if the colors' physical properties are different.

The metameric gray pair introduced in chapter 2 is shown again in figure 4.1: one sample was made with cobalt blue, yellow, and magenta, the other with ultramarine, orange, and green. A match was produced for the CIE 1931 standard observer viewing the pair under daylight. When the lighting is changed to low-wattage incandescent light, the two samples mismatch quite badly. The tristimulus XYZ signals—equivalent to LMS signals—are different for each sample.

The gray pair is one of eight metameric pairs produced for this book and shown in figure 4.2. Each pair was formulated to match under daylight for the CIE 1931 standard observer. When a low-wattage incandescent light illuminates the pairs, however, severe mismatches occur, revealed in figure 4.3. The incandescent light has a greater amount of

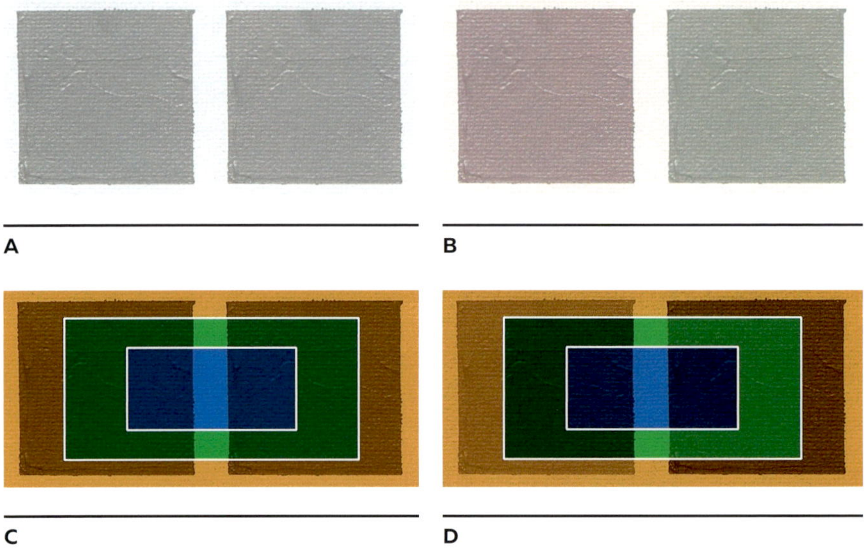

FIGURE 4.1. A gray metameric pair illuminated by (A) daylight and (B) low-wattage incandescent light. The long (orange), medium (green), and short (blue) cone-receptor images (medium and short are cropped) indicate that (C) under daylight, the LMS signals match, while (D) under incandescent light, the L and M signals are different.

FIGURE 4.2. Color pairs designed to match when illuminated with daylight. (A) The acrylic dispersion paints used for each color patch, with (B) their spectral data. Because different paints were used for each pair, their spectra are quite different. (Cobalt blue produces the high reflectance at long wavelengths. Mixtures containing arylide yellow and pyrrole orange produce different spectra compared with mixtures using arylide yellow and either pyrrole red or quinacridone magenta.)

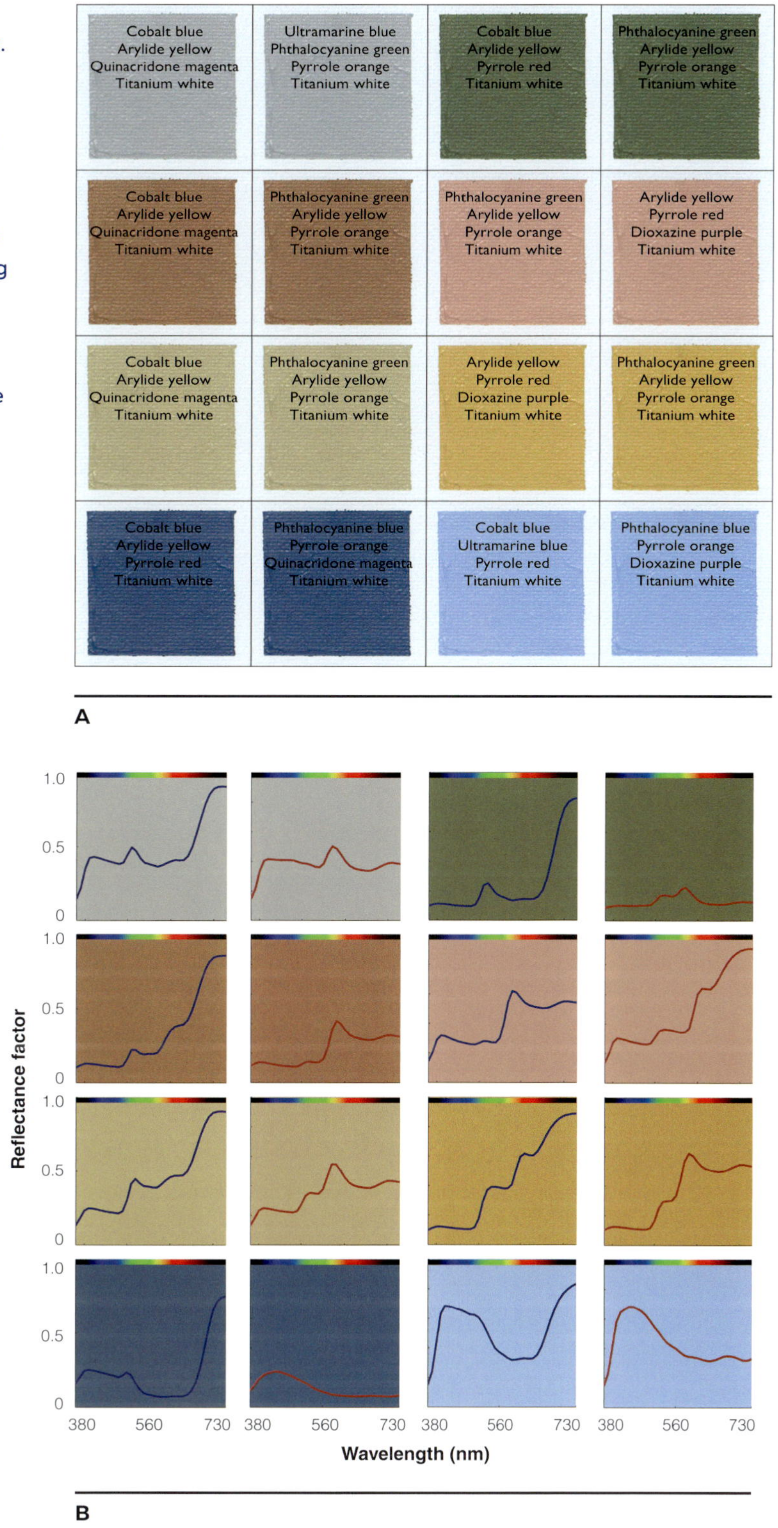

FIGURE 4.3. Eight metameric pairs (A) illuminated by daylight and viewed by the standard observer, (B) illuminated by low-wattage incandescent light and viewed by the standard observer, and (C) illuminated by daylight and viewed by the uncommon observer.

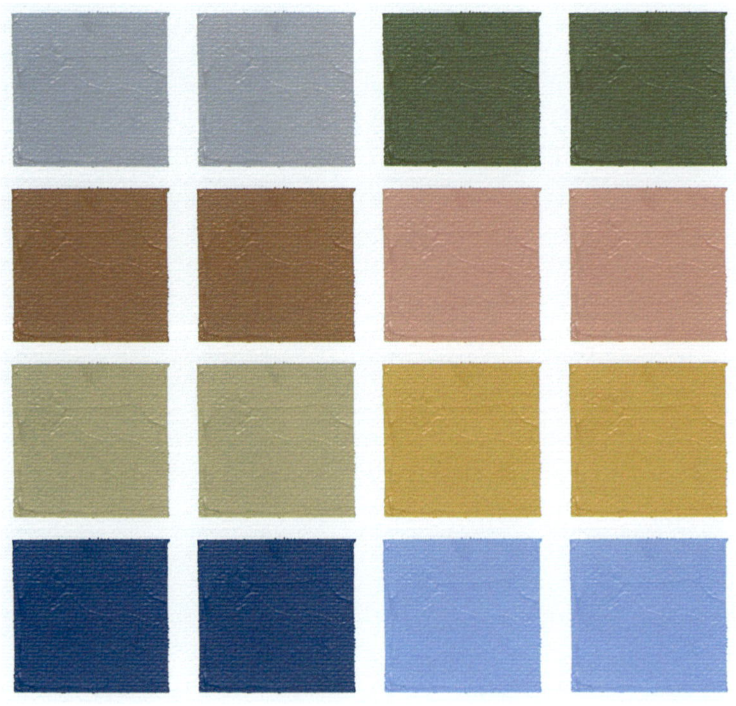

A

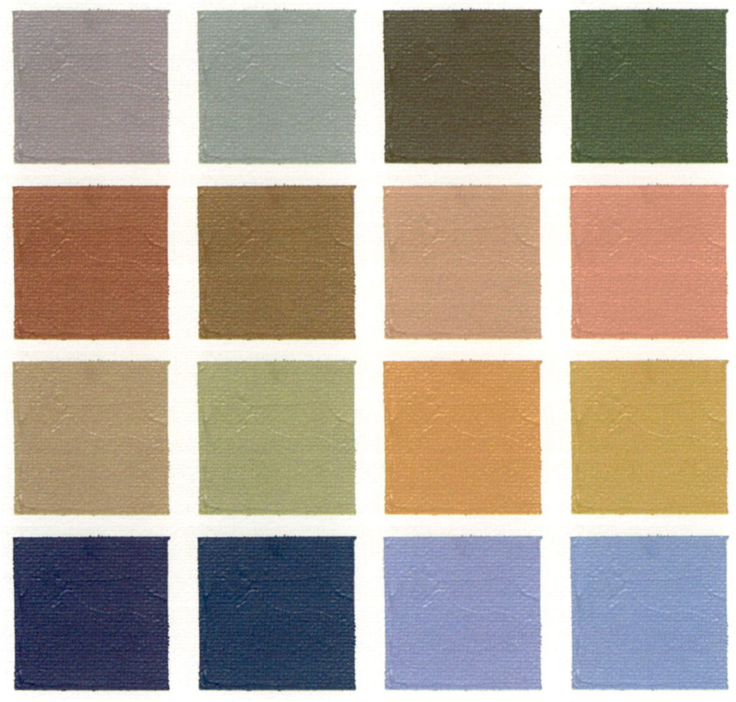

B

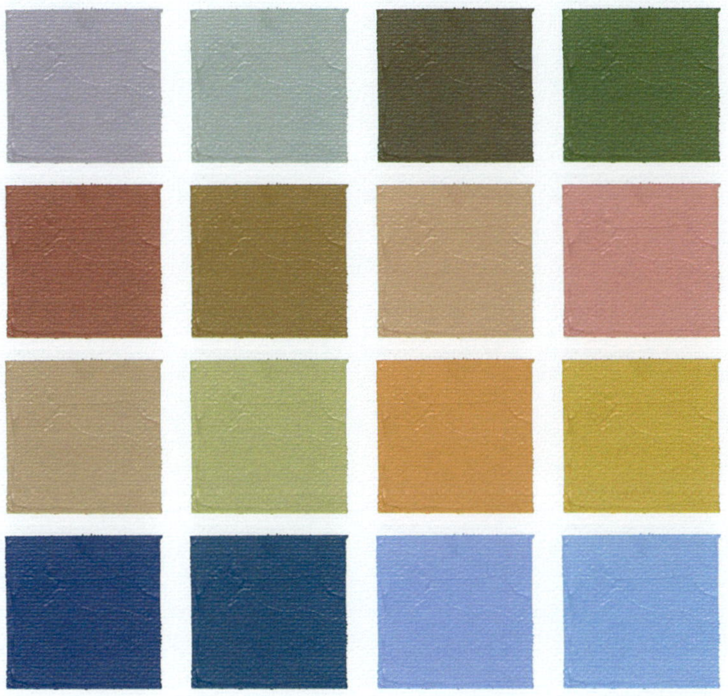

C

long-wavelength light than daylight does — the comparison is shown in figure 4.4 — and each metameric pair has large differences in spectral reflectance at longer wavelengths. When the pairs are illuminated by incandescent light, the differences in spectral reflectance are accentuated, leading to color mismatches. A color mismatch caused by changes in lighting is called illuminant metamerism.

For color-normal observers, differences in color vision are caused by the yellowing of our lens over time, by differences in the amount of macular pigment — a yellowish pigment in the center of our eye — and by differences in the spectral sensitivity of each cone type. If we know how each parameter varies from person to person and the population statistics of the variability, the range of color-normal vision can be determined. One thousand color-normal observers were calculated in this manner (Fairchild and Heckaman 2013), and the observer most different from the 1931 standard observer was selected and is referred to here as the uncommon

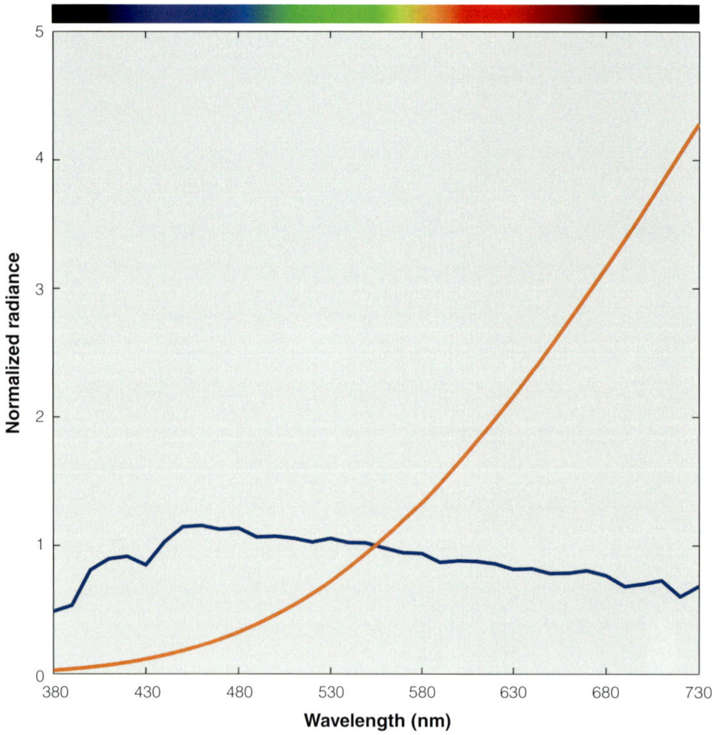

FIGURE 4.4. Spectral radiance of daylight (blue line) and a low-wattage incandescent light (orange line) normalized at 555 nm.

observer. When the uncommon observer views the eight metameric pairs, there are large mismatches, even in daylight (see figure 4.3c). This is known as observer metamerism. The degree of mismatch for any color-normal observer will range between that of the standard and uncommon observers.

Pablo Picasso's *The Tragedy* underwent conservation treatment during the early 1960s. This panel painting had had numerous paint losses owing to dimensional changes in the wooden substrate. The conservation studio was lit with natural daylight, and it is reasonable to assume that the inpainting matched the surrounding area when seen by the conservator. In the post-treatment photograph shown in figure 4.5, taken under incandescent lighting, the inpainting appeared magenta. Though the difference is less severe than in the photograph, the inpainting can be seen when the painting is displayed in an incandescent-lit gallery. Spectrophotometric

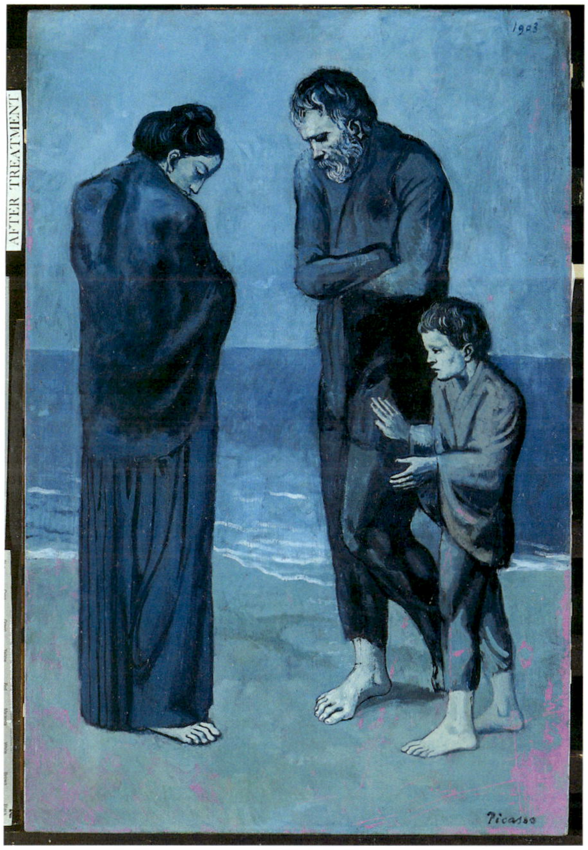

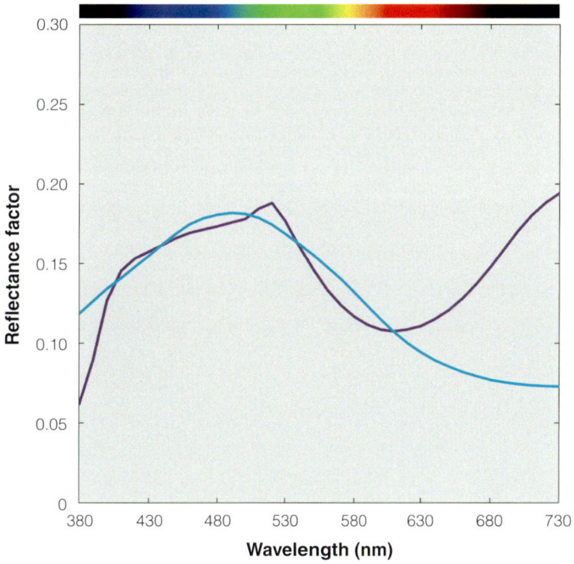

A

B

FIGURE 4.5. (A) Scanned 1963 photograph of Pablo Picasso (Spanish, 1881–1973), *The Tragedy,* 1903. Oil on wood, 105.3 × 69.1 cm (41½ × 27¼ in.). Washington, National Gallery of Art, 1963.10.196. Inpainted areas appear magenta. (B) Inpainted (purple line) and adjacent original (cyan line) passages were measured using a hand-held spectrophotometer. The spectral differences are typical of a metameric pair matching under daylight and mismatching under incandescent lighting.

measurements reveal a classic metameric pair that matches under daylight and mismatches under incandescent light, similar to the spectra of the eight metameric pairs graphed in figure 4.2. Comparing the painting's spectra to a reference library of blue pigments (Berns, Krueger, and Swicklik 2002) revealed that the conservator used ultramarine while Picasso used Prussian blue, verified using X-ray fluorescence. Blue colors in general and skies in particular are susceptible to such mismatches when the wrong colorants are selected (Staniforth 1985; Digney-Peer et al. 2012).

Color Inconstancy

Changing the lighting alters the color of most materials. We are often unaware of these changes because we expect objects to have a constant appearance. However, when we look intently, we will see that many materials change their color with lighting, in some cases dramatically. This phenomenon is known as color inconstancy. In figure 4.6, the samples forming the metameric pairs are shown illuminated by daylight and by low-wattage incandescent lighting. Most of these samples are color inconstant, and in some cases the color changes are dramatic. The specific color changes depend on the spectral properties of the sample (Berns, Billmeyer, and Sacher 1985) and the type of lighting.

The CIELAB color target introduced in chapter 3 contains twelve samples with different hues. As seen in figure 4.7, these samples, too, exhibit color inconstancy in these two types of lighting. A palette of vivid colors is always more color inconstant than a palette composed mainly of earth colorants, such as those used by the old masters.

Because of color inconstancy, paintings can appear quite different under different types of lighting, depending on the artist's palette. An example is shown in figure 4.8, Vincent van Gogh's *Bedroom* in two types of lighting. Van Gogh used cobalt blue in the doors and walls of the painting (Geldof, Megens, and Salvant 2013). Under incandescent lighting, such as from the PAR spotlights often used in museums, the cobalt blue becomes reddish. The other blues in the painting are Prussian blue and ultramarine. If Van Gogh had used more Prussian blue, substituting it for all the cobalt blue, the painting would be more color constant.

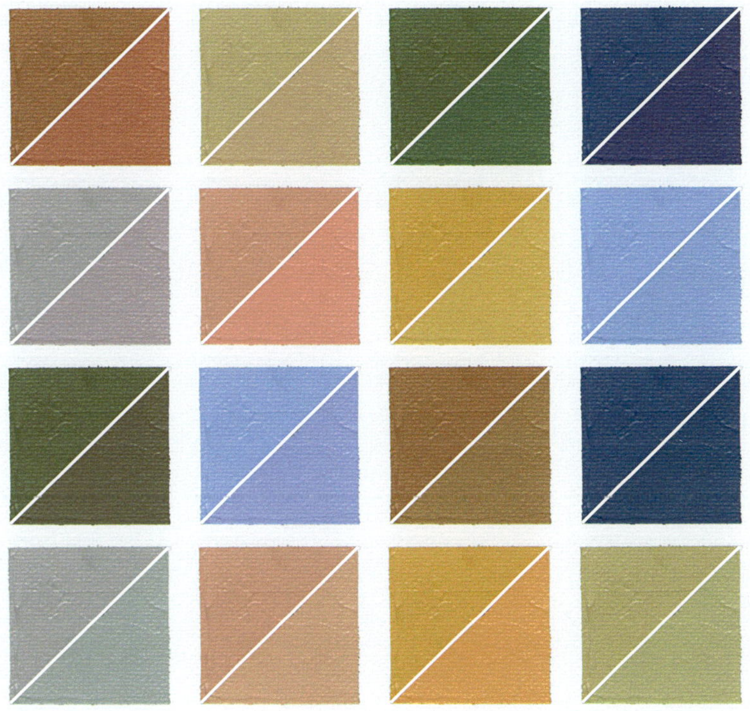

FIGURE 4.6. The shuffled colors of the sixteen samples forming the eight metameric pairs when illuminated by daylight (upper left triangle) or low-wattage incandescent lighting (lower right triangle).

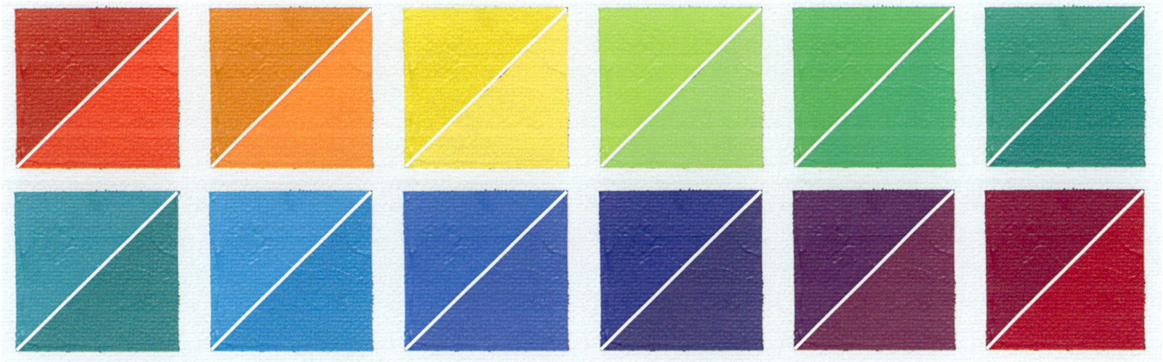

FIGURE 4.7. Twelve samples varying in hue illuminated by daylight (upper left triangle) or low-wattage incandescent lighting (lower right triangle).

FIGURE 4.8.
Vincent van Gogh (Dutch, 1853–1890), *Bedroom,* 1889. Oil on canvas, 73.7 × 93 cm (29 × 36⅝ in.). Art Institute of Chicago, 1926.417. (A) Actual painting illuminated by either daylight (outer frame) or incandescent light (center).
(B) Simulated painting with Prussian blue replacing cobalt blue, illuminated by either daylight (outer frame) or incandescent light (center).

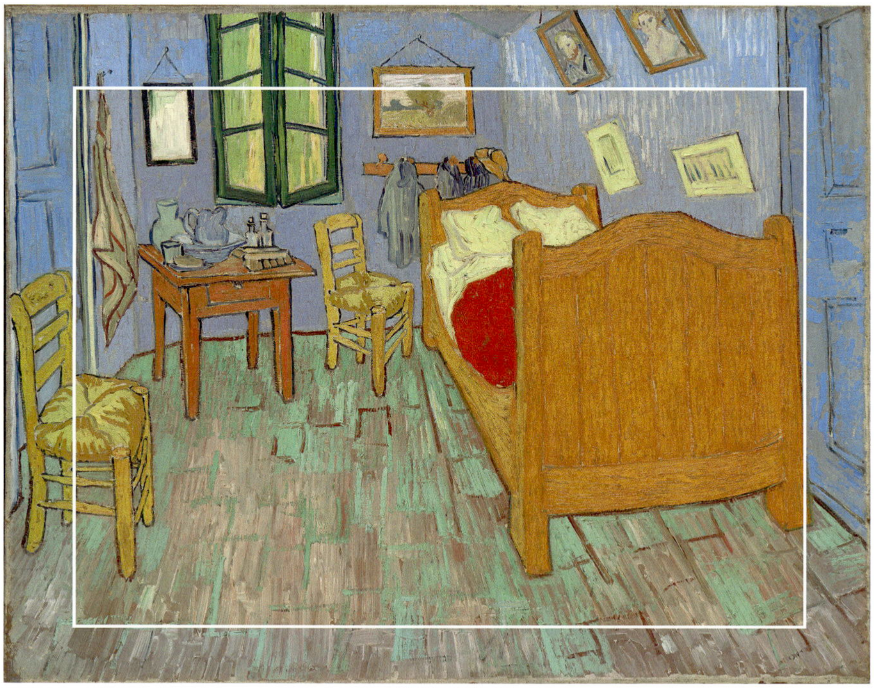

A

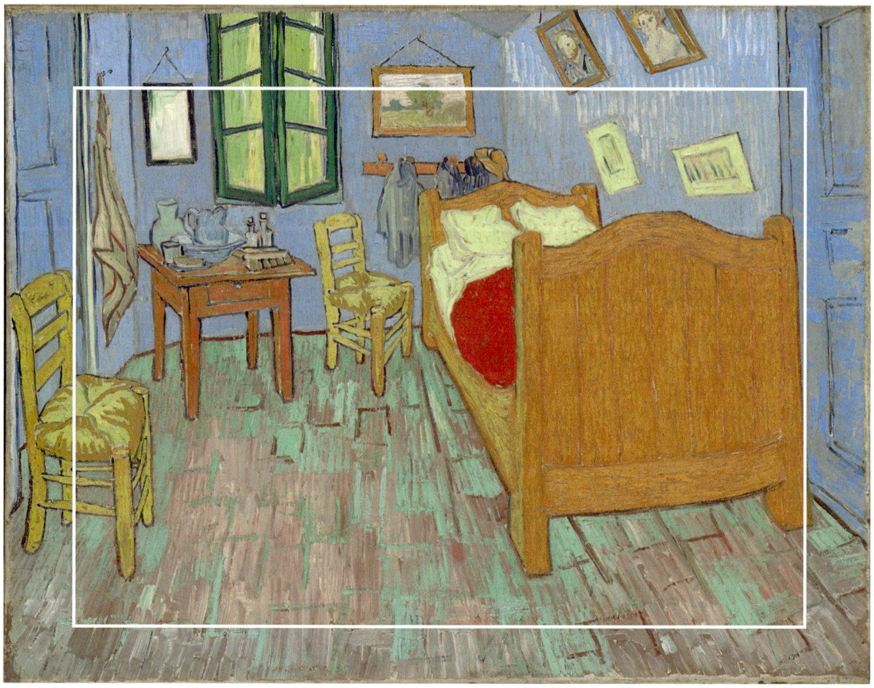

B

Summary

Metamerism refers to two or more objects, whereas color inconstancy refers to a single object. Two objects that match under a single viewing condition but have dissimilar spectral reflectance are metameric. When a change in lighting results in a color mismatch, the effect is illuminant metamerism. When a change in observer results in a color mismatch, the effect is observer metamerism. Metamerism is an important consideration when selecting colorants for imitative inpainting. A single object is color inconstant when its color alters with changes in lighting. Most materials and paintings are color inconstant, which can sometimes produce unwanted shifts in color under different light sources. The distinction between metamerism and color inconstancy is important. When people refer to a single object as being metameric, they mean to say that the object is color inconstant.

References

Berns, R. S., J. F. W. Billmeyer, and R. S. Sacher. 1985. "Methods for Generating Spectral Reflectance Functions Leading to Color-Constant Properties." *Color Research and Application* 10: 73–83.

Berns, R. S., J. Krueger, and M. Swicklik. 2002. "Multiple Pigment Selection for Inpainting Using Visible Reflectance Spectrophotometry." *Studies in Conservation* 47: 46–61.

Digney-Peer, S., K. Thomas, R. Perry, J. Townsend, and S. Gritt. 2012. "The Imitative Retouching of Easel Paintings." In *The Conservation of Easel Paintings,* edited by J. H. Stoner and R. Rushfield, 607–34. New York: Routledge.

Fairchild, M. D., and R. L. Heckaman. 2013. "Metameric Observers: A Monte Carlo Approach." In *21st Color and Imaging Conference,* 185–90. Albuquerque, NM: Society of Imaging Science and Technology.

Geldof, M., L. Megens, and J. Salvant. 2013. "Van Gogh's Palette in Arles, Saint-Rémy and Auvers-sur-Oise." In *Van Gogh's Studio Practice,* edited by M. Vellekoop, M. Geldof, E. Hendriks, L. Jansen, and A. de Tagle, 238–55. New Haven, CT: Yale University Press.

Staniforth, S. 1985. "Retouching and Colour Matching: The Restorer and Metamerism." *Studies in Conservation* 30: 101–11.

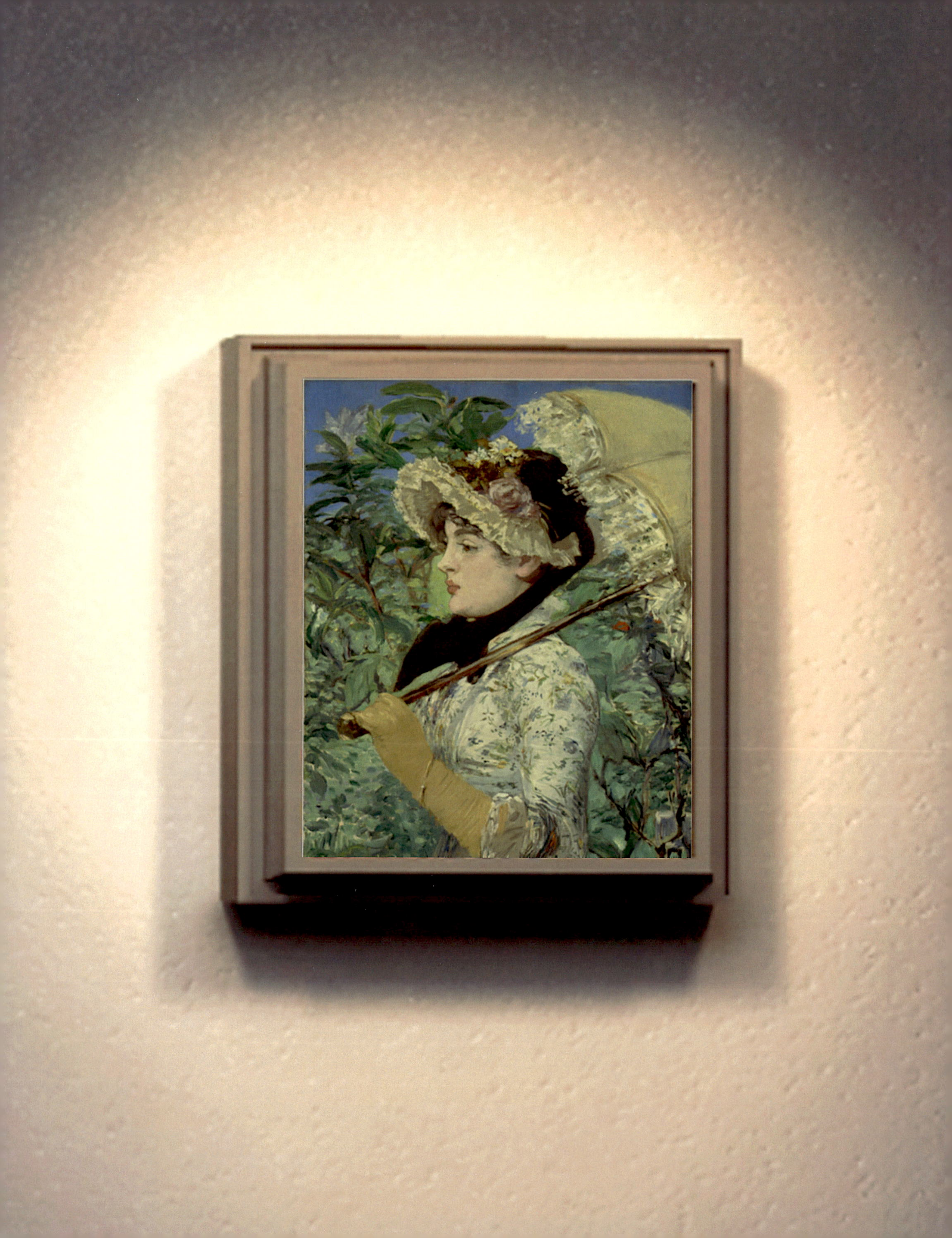

DISPLAYING ARTWORK

5

We look at paintings not in isolation but rather within a space such as a gallery or studio. A painting's appearance depends on many aspects of that space, among them ambient lighting, direct lighting, and the color of the ceiling, walls, and flooring. Important properties of lighting include energy efficiency, spectral radiance, color, intensity, and geometry. Short-wavelength visible and UV radiation can cause light-induced damage and should be minimized in gallery lights. The intensity of direct illumination should also be controlled to minimize potential light-induced damage.

The appearance of Édouard Manet's *Le Printemps (Jeanne Demarsy)* depends on ambient lighting, direct lighting, and the color of the ceiling, walls, and flooring (detail, fig. 5.6b).

Because most materials, including those used in artwork, are color inconstant, selecting a light's color is an important decision. "Cooler" lights tend to increase the vividness of blues and greens and decrease the vividness of reds and yellows; "warmer" lights tend toward the opposite. We think of an object's "true" color as the color we observe when the object is illuminated by daylight and incandescent light.

Today, there are many technologies that produce light, particularly solid state. These lights can both produce true colors and change colors by increasing or decreasing vividness, providing new opportunities in museum lighting. The geometry of lighting, whether diffuse or focused, masks or accentuates surface topography such as brushstrokes, impasto, and texture of the painting support. The color of the ceiling, walls, and flooring has an impact on the amount of diffuse light in the viewing space. Lighter surfaces reflect more light while darker surfaces absorb it. All of these factors combine to influence our comfort and ability to see artwork within a gallery space.

Illumination Level

Our visual system enables us to see over an extremely wide range of light intensities, from starlight to full sunlight. This light adaptation results from the change in size of our pupils and from adjustments in our receptors and our brain. We have two different types of sensors, rods and cones, which are optimized for dark conditions (rods) and for a wide range of lighting (cones).

Light intensity was first defined numerically in the early twentieth century through experiments that were performed to measure visibility, defined as the propensity of a light source to produce a visual response (ISO 2005). Observers adjusted individual wavelength radiances until they were equally visible. The radiance data were used to produce a visibility function, V_λ (pronounced *vee-lambda*), shown in figure 5.1. Today this is called the "photopic luminous efficiency function."

Red and green lasers were introduced in chapter 1. In figure 5.2 their radiances are adjusted to produce the same visibility. Because our sensitivity to visibility is lower at 630 nm than at 530 nm, the red laser must have over three times the amount of radiance of the green laser. Thus, looking at spectra cannot tell us how bright or dim a light will appear.

Any lighting metric that includes "lum" (from the Latin *lumen,* meaning light), such as lumen, illuminance, and luminance, accounts for the way visibility differs across wavelength, that is, incorporates V_λ in calculation. Illuminance is defined as the amount of light that falls upon a surface, often measured using a light meter, while luminance is defined as the amount of light being generated by the source. There are two common units for illuminance: foot-candles (fc) and lux (lx). Multiplying foot-candles by 10 gives an approximate conversion to lux.[1]

An important aspect of lighting is its energy efficiency, termed "luminous efficacy." A 100-watt incandescent bulb (110 volts) produces about 1700 lumens. A warm-white compact fluorescent light (CFL) bulb

FIGURE 5.1. Visibility as a function of wavelength, V_λ, defined as the photopic luminous efficiency function.

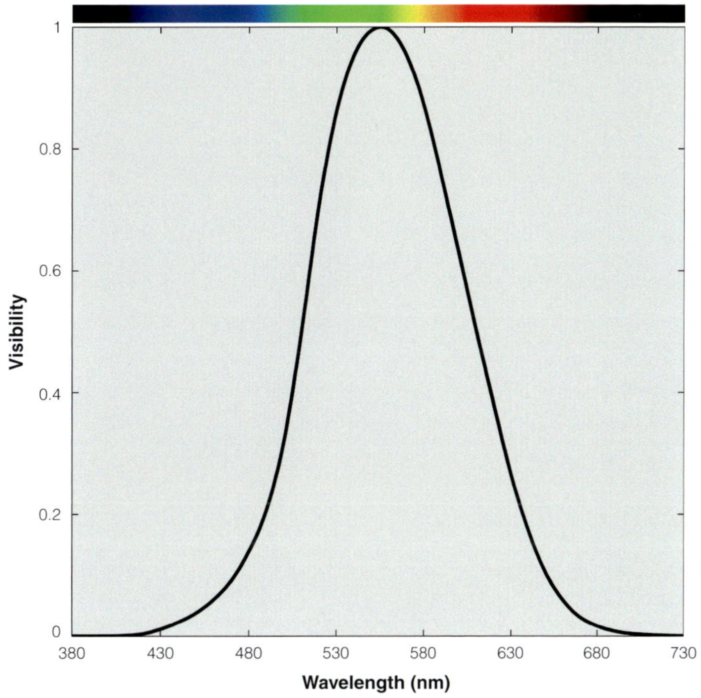

FIGURE 5.2. Spectral radiance of red and green lasers producing identical visibility. The visibility curve is also plotted, showing that much of the red light emits radiance where visibility is low.

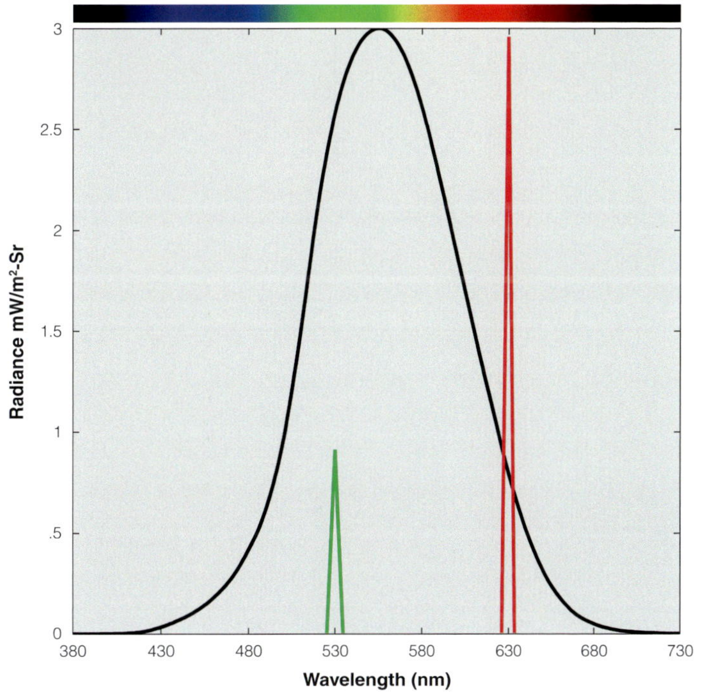

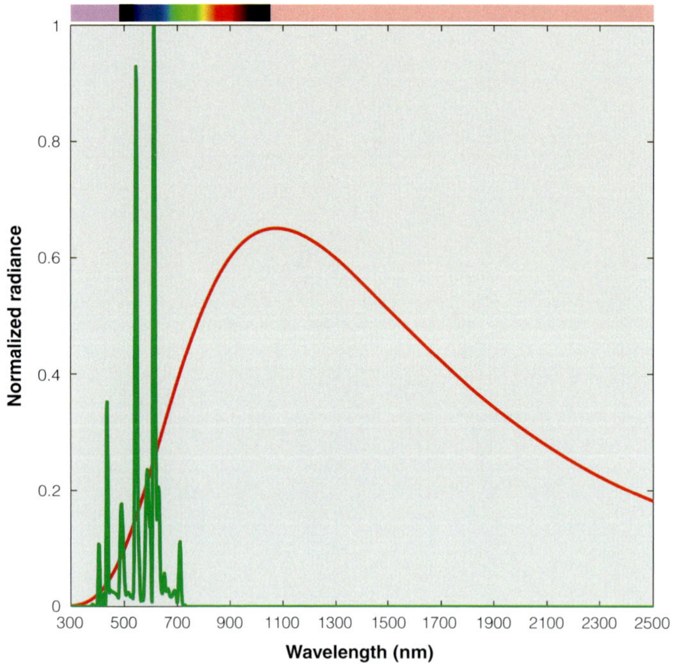

FIGURE 5.3. Spectral radiance of PAR incandescent (red line) and warm-white CFL (green line) lights normalized to equal luminance. Beyond the visible range, the incandescent lamp produces considerable infrared radiance. Note that the wavelength range includes ultraviolet, visible, and infrared regions from left to right.

requires about 25 watts to produce the same 1700 lumens. Thus, the CFL is four times more energy efficient in producing light. Another aspect of energy efficiency, illustrated in figure 5.3, is the amount of cooling that is required to compensate for the large amount of infrared radiation generated by incandescent lighting. The poor luminous efficacy and appreciable infrared radiation are the reasons why incandescent lighting is being phased out for general lighting.

Illuminance levels for paintings fall into a wide range. An artist outdoors in full sunlight may work at levels of up to 150,000 lx, whereas in the studio at night the levels are closer to 500 lx. Museum lighting can vary considerably in illuminance, depending on the artworks' vulnerability to light-induced damage. Minimum levels are around 50 lx. Reducing illuminance worsens visual acuity and color discrimination as well as the range of perceived colors. In figure 5.4 these effects are simulated for Henri de Toulouse-Lautrec's *The Model Resting,* lit with illuminances of 1500 lx and 50 lx.

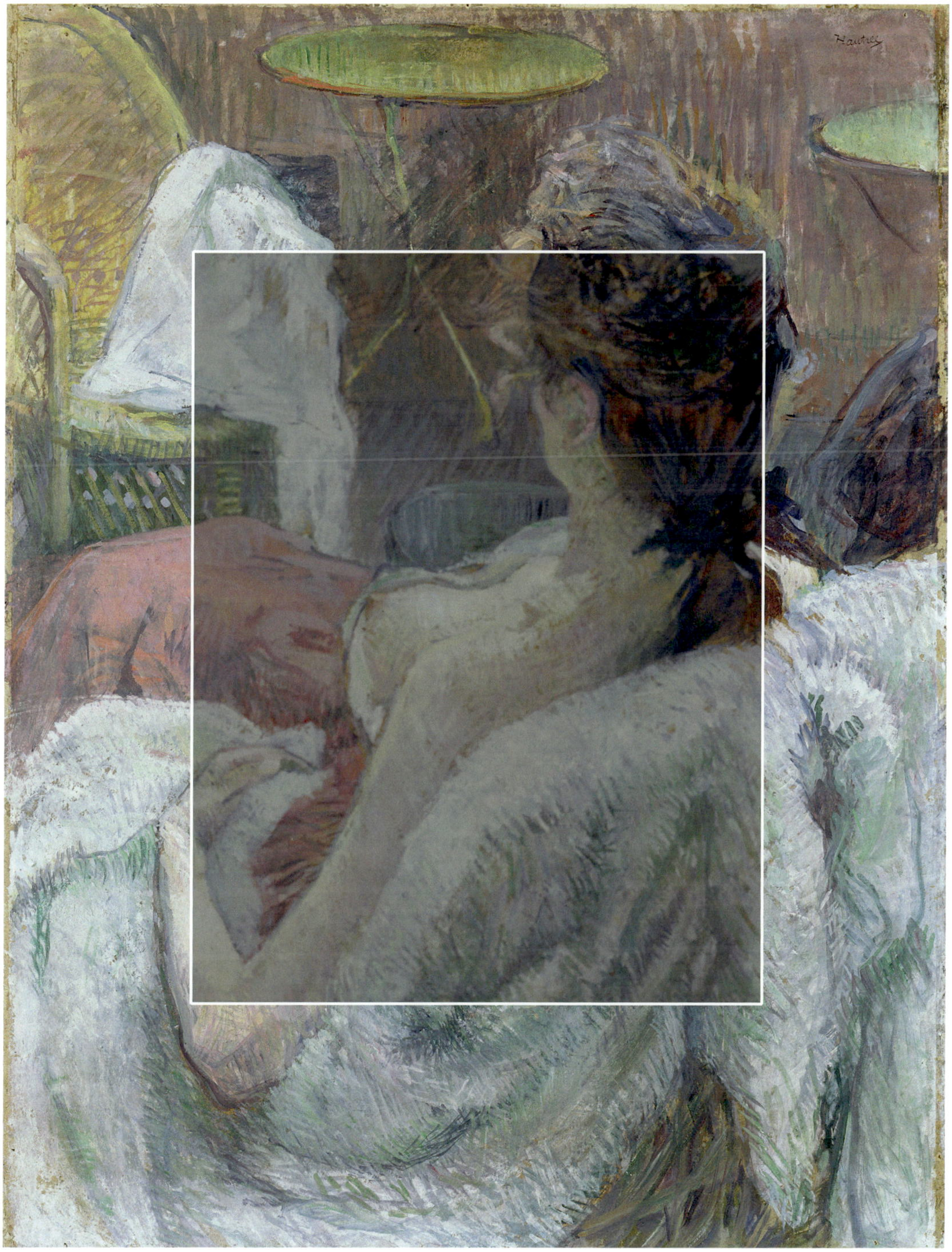

FIGURE 5.4. Henri de Toulouse-Lautrec (French, 1864–1901), *The Model Resting*. Tempera or casein with oil on cardboard, 65.4 × 49.2 cm (25¾ × 19⅜ in.). Los Angeles, J. Paul Getty Museum, 84.PC.39. The demarcated areas show how the painting would look if lit at 1500 lx (edge) and at 50 lx (center).

Illumination Color

The majority of lighting designed for general illumination appears white. However, when we look deliberately at "white" lighting, we see that many sources have a slight hue, often bluish or reddish. This is why consumer lighting is classified as cool (bluish), white (neutral), or warm (reddish). For greater specificity, the term "correlated color temperature" (CCT) is used having units of Kelvin (K = C° + 273.15). Warm lighting can have a 3000 K CCT while a cool light can have a 6500 K CCT.[2] The PAR incandescent lights commonly used in museums have about a 2700 K CCT, while north-sky light, often used in conservation studios, is near 7500 K.

CCT originates with a material that absorbs all incident radiation, called a blackbody because at room temperature it appears black. The absorption heats the blackbody, which in turn causes it to emit its own radiation. The relationship between a blackbody's temperature and radiation is determinate (Planck 1901), and at extremely high temperatures the radiation produces light. Blackbody radiators at 3000 K and 6500 K are graphed in figure 5.5. With an increase in temperature, there is a large increase in total energy and a shift in the maximum radiance to shorter wavelengths. By replotting these spectra normalized to equal luminance, we see that the majority of the radiation produced at 3000 K is in the infrared. At 6500 K there is more ultraviolet radiation. These spectra can also be normalized to the same value at 555 nm, the center of the visibility function, demonstrating that lower color temperatures produce redder lighting and higher color temperatures produce bluer lighting. Strictly speaking, the term "color temperature" is reserved for blackbodies; all other sources are described using correlated color temperature. A source with a specific CCT matches the color of the light produced by a blackbody at the same temperature.

Historically, museums have had limited lighting choices, generally either incandescent lamps ranging in CCT from about 2000 K (when dimmed) to 3200 K, or daylight ranging in CCT from about 5000 K to 7500 K. Because most paintings are color inconstant, these differences have a profound effect on a painting's colors. The degree of difference depends on the artist's palette, as shown for Vincent van Gogh's *Bedroom* in chapter 4.

The effects of CCT on a painting's appearance are demonstrated in figure 5.6. Édouard Manet's *Le Printemps (Jeanne Demarsy)* was duplicated computationally, using a palette—chosen to minimize color

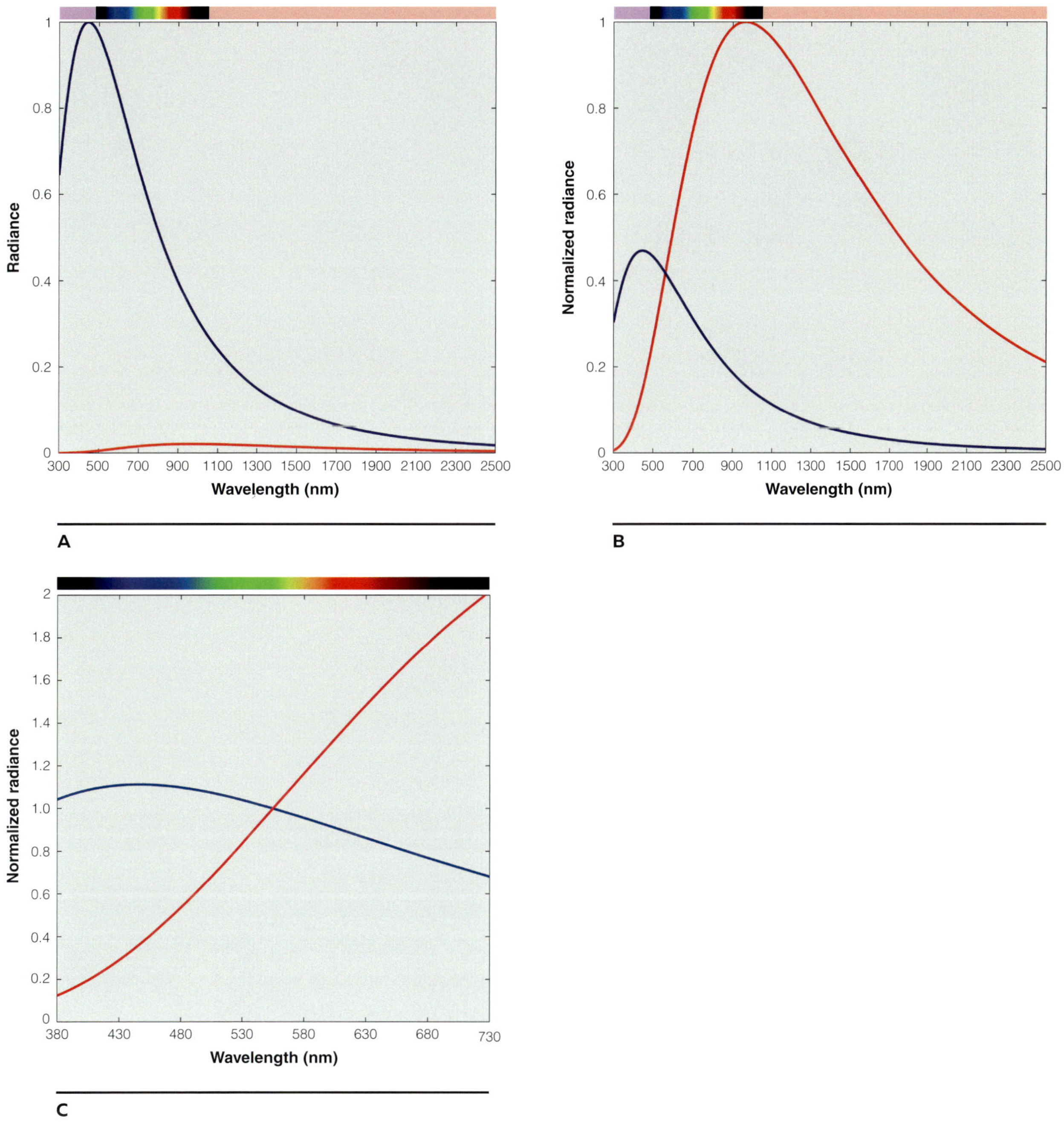

FIGURE 5.5. (A) Spectral radiance of blackbody radiators at 3000 K (red line) and 6500 K (blue line). The spectra are normalized to have (B) equal luminance and (C) equal radiance at 555 nm.

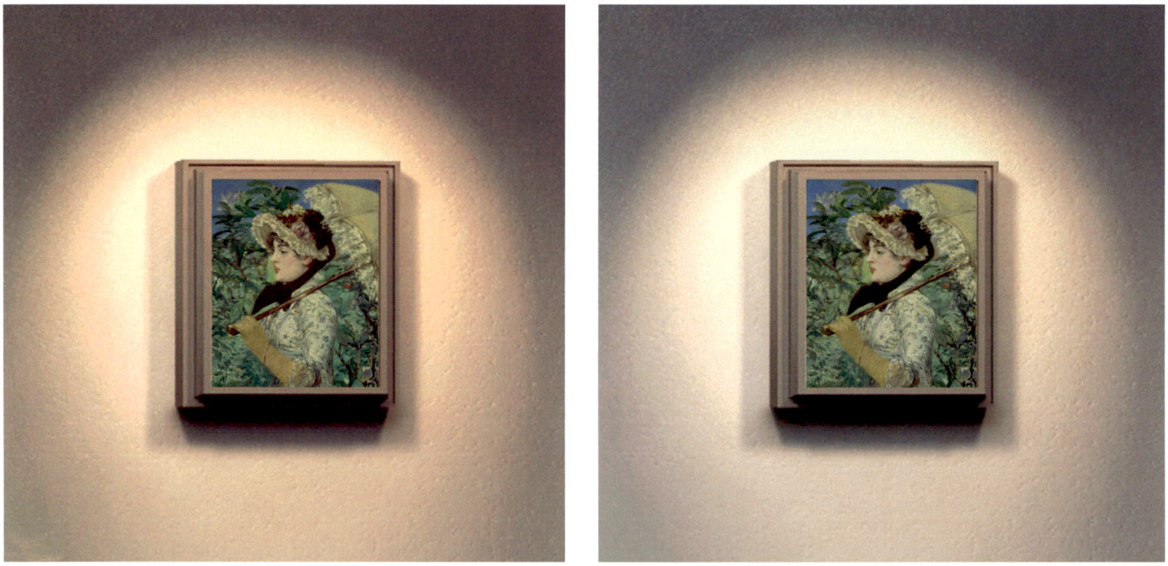

A B

FIGURE 5.6. Duplicate of Édouard Manet (French, 1832–1883), *Le Printemps (Jeanne Demarsy)*, 1881. Oil on canvas, 74 × 51.4 cm (29⅛ × 20¼ in.). Los Angeles, J. Paul Getty Museum, 2014.62. Prussian blue, arylide yellow, quinacridone magenta, and titanium white were used in the duplicate, illuminated here with blackbody radiation at (A) 2000 K, (B) 3000 K, (C) 4000 K, and (D) 5000 K.

inconstancy in the blue sky—containing Prussian blue, arylide yellow, quinacridone magenta, and titanium white. The duplicate painting was then rendered for 2000, 3000, 4000, and 5000 K blackbody radiation. The skin tones are warmer at lower color temperatures, even after cognitively discounting the color of the illumination reflected from the wall.

Experiments have been performed where paintings are placed in light booths and observers can select a preferred CCT. When paintings were spot-illuminated against a black background, preferences were near 5500 K (Nascimento and Masuda 2014). When paintings were illuminated diffusely against a white background, the preferred CCT depended on the painting, with variability among the observers ranging from 2600 K to 6400 K (Liu et al. 2013). A separate study was performed in a modern-art gallery comparing 2700 K and 3000 K illumination (Miller and Rosenfeld 2012). The lighting designers preferred the higher CCT, as it increased the vividness of blue skies and other blue passages. The

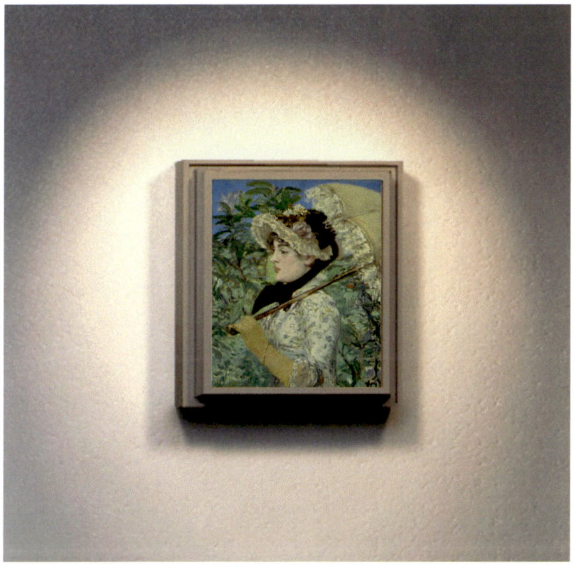

C

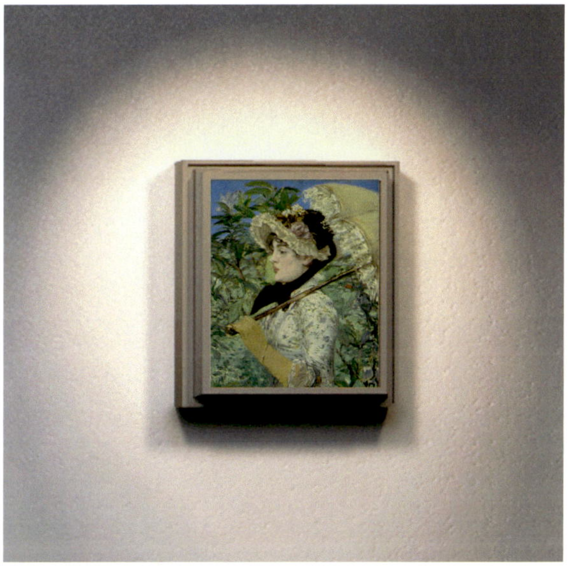

D

curatorial staff and museum director, however, disliked the increase in vividness and found the relatively bluer color of the lighting distracting. As these experiments reveal, there is no single optimal CCT that would be universally acceptable when illuminating paintings.

Color Rendering

The terms "artificial" and "natural" are often used to distinguish light sources. If candlelight is natural, should an incandescent light bulb also be considered natural, as the physical process of incandescence is identical? Or is a light bulb artificial? The term "artificial" has a negative connotation as being unnatural. The lighting industry uses the term "electric lighting" instead, since all sources that have an on-and-off switch use electricity.

Over the decades electric lights have been developed that produce light other than through incandescence. These lights, using fluorescent-tube, metal-halide, and solid-state technologies (Khanh et al. 2015), among others, produce a range of spectral characteristics and CCTs. A three-band light, representative of an RGB-type solid-state light, was designed that matches the color of 4500 K blackbody radiation (Berns 2011). Both spectra are graphed in figure 5.7. Then the CIELAB target introduced in chapter 3 was rendered for both "lights." As seen in figure 5.8, the neutrals are identical in color, verifying the equality of the lights' colors. However, the three-band light produced large color distortions; two of the blues became a vivid magenta, while the reddish yellows became greenish. Unsurprisingly, this source was designed to maximize color distortions.

The need to quantify the extent of distortion as compared with our common viewing experience arose with the development of fluorescent lighting, which has different spectral properties from those of daylight and incandescent light. The term that was subsequently adopted, "color rendering," is defined as the effect of a light source on the color appearance of objects in comparison with their color appearance under a reference light source (ASTM 2012). Research began in the 1950s (Nickerson 1960), and in 1965 the CIE recommended a method and metric to quantify the color rendering of lighting (CIE 1995). The metric, known as the color rendering index (CRI), is universally used within the lighting industry.

CRI is calculated using reference illuminants and samples. The reference illuminants render colors that match our expectations of an object's "true color." Two classes of reference illuminants were selected for this purpose and graphed in figure 5.9: CIE daylight representing daylight at CCTs of 5000 K and above, and blackbody radiation representing incandescent light, sunrise, and sunset at CCTs below 5000 K.

The set of reference samples shown in figures 5.10–12 was selected from the Munsell system: eight hues having similar lightness and chroma. For each reference sample, a total color difference is calculated comparing the sample's color under a test source with its color under a reference illuminant, the reference illuminant having been selected to match the test source in its CCT. Comparing lights with the same CCT factors out color inconstancy. The color differences are combined, resulting in CRI. A test light has a CRI of 100 when it renders the eight colors identically compared with the reference illuminant. The index was developed so that warm-white fluorescent light, known to have poor rendering, had a CRI of 50. A CRI of at least 85 has been recommended for museums (Cuttle 2007).

The CIE method of quantifying color rendering has been criticized, both for its method of computation and its definition of color rendering (Rea and Freyssinier 2008, 2010; Davis and Ohno 2010; Burgt and Kemenade 2010). The eight samples are a particular problem because of their low chroma; as a consequence, the index has low sensitivity. This is

FIGURE 5.7. Spectral radiance normalized to equal luminance of a 4500 K blackbody radiator (orange line) and of a solid-state light (blue line) designed to produce large color distortions while also having a 4500 K CCT.

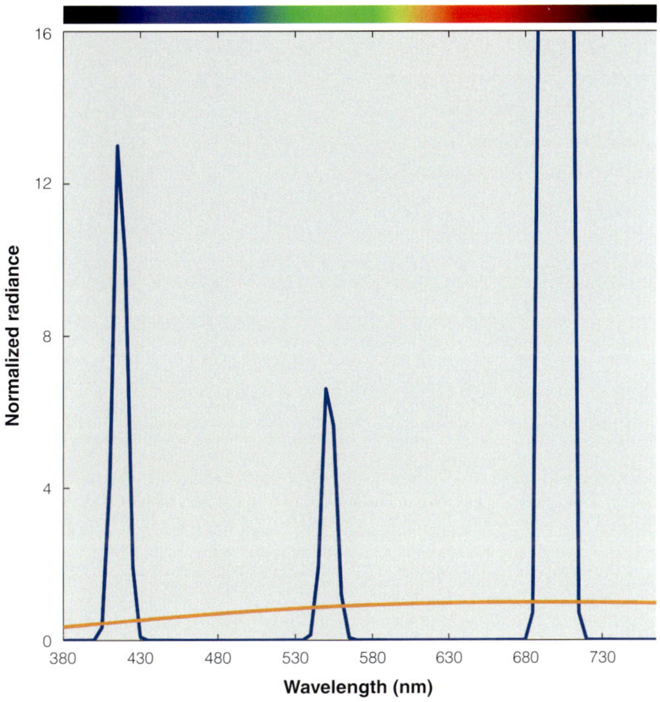

FIGURE 5.8. CIELAB target of colors rendered for 4500 K blackbody radiation (top triangle) and a three-band light designed to produce large color distortions while also having a 4500 K CCT (bottom triangle).

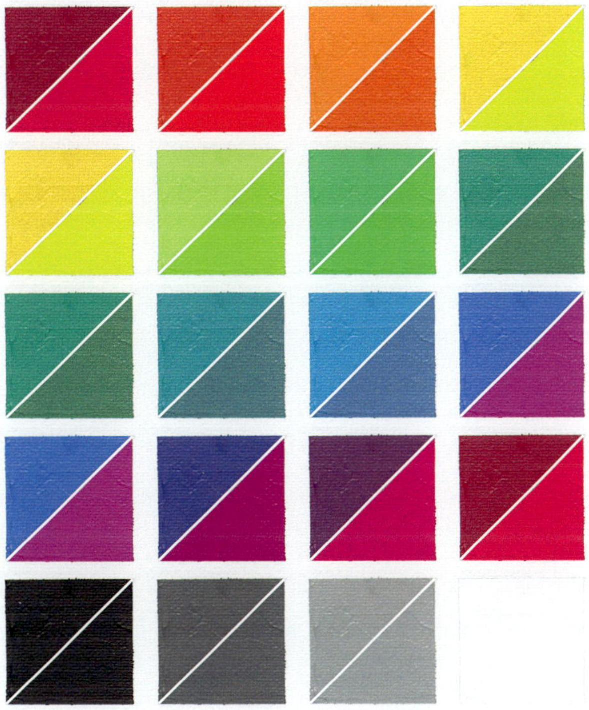

FIGURE 5.9. Spectral radiances normalized to equal radiance at 555 nm of (A) CIE daylight ranging from 5000 K (greenish blue line) to 7500 K (blue line), and (B) blackbody radiation ranging from 2000 K (red line) to 4500 K (yellow line), all in 500 K increments.

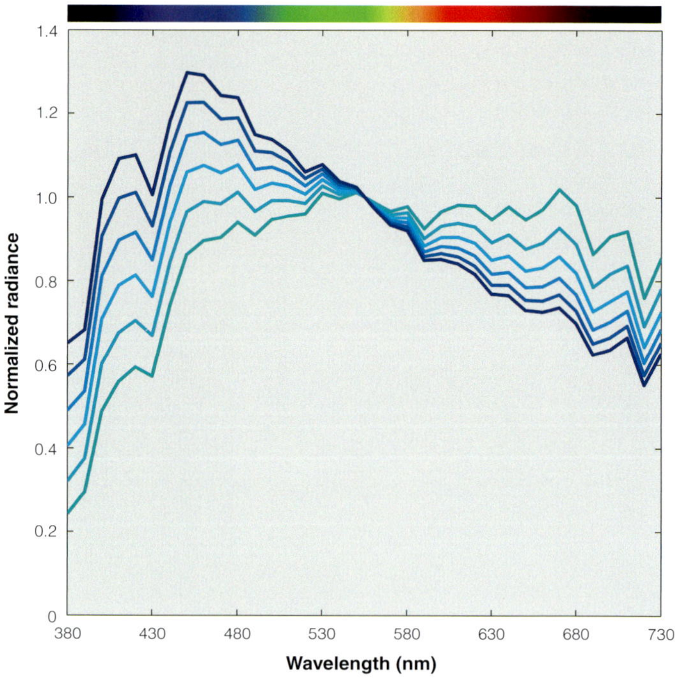

A

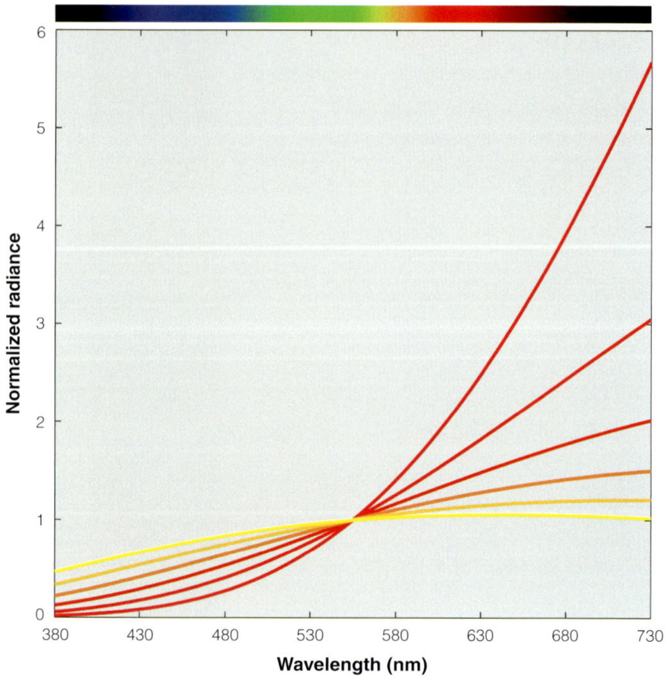

B

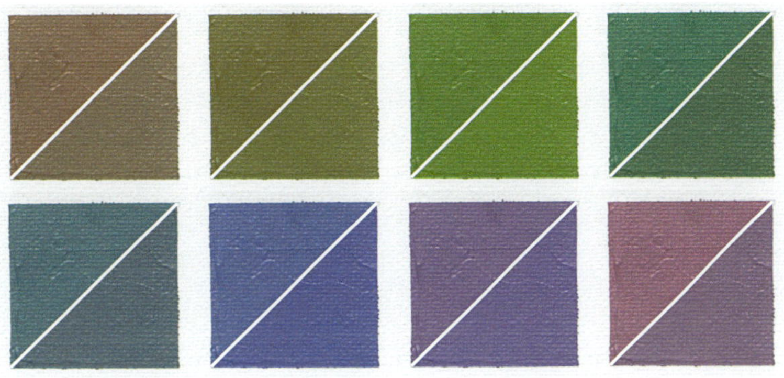

FIGURE 5.10. Eight CIE color-rendering index samples rendered for 2900 K blackbody radiation (top triangle) and warm-white fluorescent with a 2900 K CCT (bottom triangle).

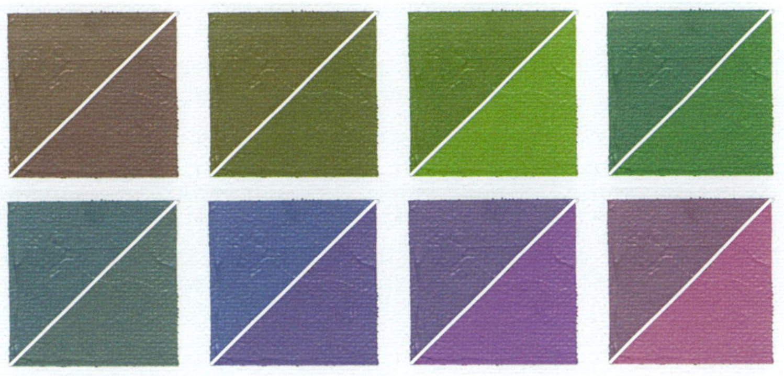

FIGURE 5.11. Eight CIE color-rendering index samples rendered for 4500 K blackbody radiation (top triangle) and the three-band light designed to produce large color distortions while also having a 4500 K CCT (bottom triangle).

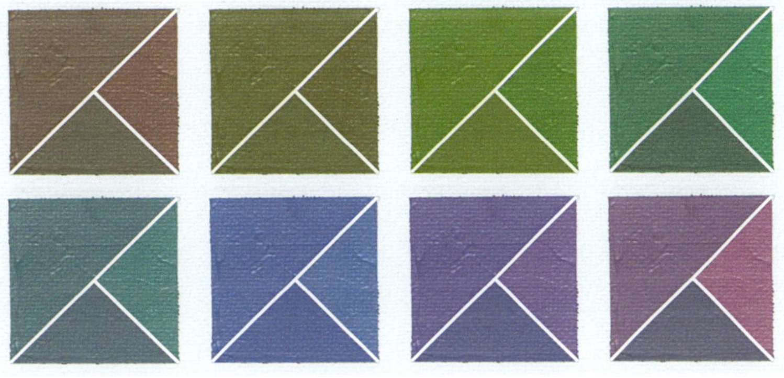

FIGURE 5.12. Eight CIE color-rendering index samples rendered for 6500 K CIE daylight (top left triangle) and a solid-state light designed to increase chroma (right triangle) or decrease chroma (bottom triangle) while also having a 6500 K CCT.

demonstrated in figure 5.11 for the distorting three-band light. The color changes for the CRI samples are small in comparison to those shown for the CIELAB target (see figure 5.8).

Since CRI is based on the calculation of a total color difference, only the magnitude of change is evaluated, not the direction. Two solid-state lights were used to illuminate the eight CRI samples, shown in figure 5.12. One was designed to increase chroma, the other to decrease it, while maintaining hue and lightness (Berns 2011; Žukauskas et al. 2012). Both sources have CRIs near 50, yet one may be preferred despite having a low CRI. It has been suggested that two indices are necessary to define color rendering (Judd 1967; Rea and Freyssinier 2008, 2010; David et al. 2015), the first an updated CRI that could indicate the "naturalness" of a source and the second measuring the change in color gamut, that is, indices of fidelity and preference. A source that increases vividness might be preferred over a source with high fidelity.

If CRI is used as the sole selection criterion for museum lighting, sources that deliberately increase vividness or dullness are excluded from consideration. Is this the correct criterion? Works on paper often have limited lightfastness and are displayed at or below 50 lx. Color gamut is reduced (see figure 5.4), particularly in comparison to the light levels most likely used by the artist. A light that increases vividness could help compensate for this reduction. Such a light was fabricated and used to counteract the fading of several natural history specimens (Viénot, Coron, and Lavédrine 2011). Suppose a Renaissance artist expected a painting to be viewed in a church where the dominant light source was candlelight. Viewing the painting in a museum at a much higher illuminance would increase the range of lightness and chromatic intensity well beyond the artist's intent. A dulling light could help compensate for these increases.

Light-Induced Damage

Many paintings and drawings are vulnerable to light-induced damage. This has resulted in the recommendation to set illuminance levels at no greater than 50 lx for highly or moderately vulnerable artwork and 150–200 lx for minimally vulnerable artwork, and to filter radiation for wavelengths shorter than 400 nm and, if practical, IR radiation (CIE 2004; DiLaura et al. 2011; Druzik and Michalski 2012). Illuminance levels can be higher for materials that are not vulnerable to light damage. There is also a preference for low CCT sources because the amount of harmful short-wavelength visible and UV radiation decreases as CCT reduces. Neither the recommendations nor the low CCT preference eliminates the potential for damage, but they should at the least minimize it. Microfading testing can be performed in an inconspicuous location on objects suspected of being susceptible to light-induced damage (Whitmore, Pan, and Bailie 1999), and the results can be useful for risk assessment (Michalski 1992; Druzik and Michalski 2012).

In practice, it may be necessary to relax these maximum levels. In the case of old master paintings, 150–200 lx may be too low to see dark passages. At 50 lx elderly patrons may not be able to see the artwork clearly or read didactic information.

The majority of today's artist materials have excellent lightfastness, but this has not always been the case. For example, lake pigments are produced by precipitating an aqueous solution of soluble organic dye mixed with an insoluble substrate such as alum, lead acetate, tin, or calcium. Depending on the specific dye and substrate, these pigments can range from lightfast to fugitive. Vincent van Gogh's Arles palette included eosin ("geranium") and cochineal ("carmine") lakes (Geldof, de Keijzer, et al. 2013; Geldof, Megens, and Salvant 2013). When mixed with blue and white, these lake pigments produced the vivid purples that were critical to Van Gogh's working method: he often sought to include complementary color pairs of purple and yellow, blue and orange, and red and green, the complements derived from Blanc's chromatic rose (Dijk 2013). In a letter to his brother Theo about the Van Gogh Museum's *Bedroom*, shown several times in this book, Van Gogh wrote:

> *The walls are of a pale violet. The floor—is of red tiles.*
> *The bedstead and the chairs are fresh butter yellow.*
> *The sheet and the pillows very bright lemon green.*
> *The bedspread scarlet red.*
> *The window green.*
> *The dressing table orange, the basin blue.*
> *The doors lilac.*[3]

Van Gogh's description of the painting does not match its current condition, in which the walls and doors are blue, not violet and lilac, as a result of these two lake pigments fading (Berg et al. 2006; Geldof, de Keijzer, et al. 2013). Van Gogh was aware of the instability of the paints, though it is doubtful he anticipated that these lake pigments would fade completely. In a letter to Theo about Guillaume Charles Tasset, who

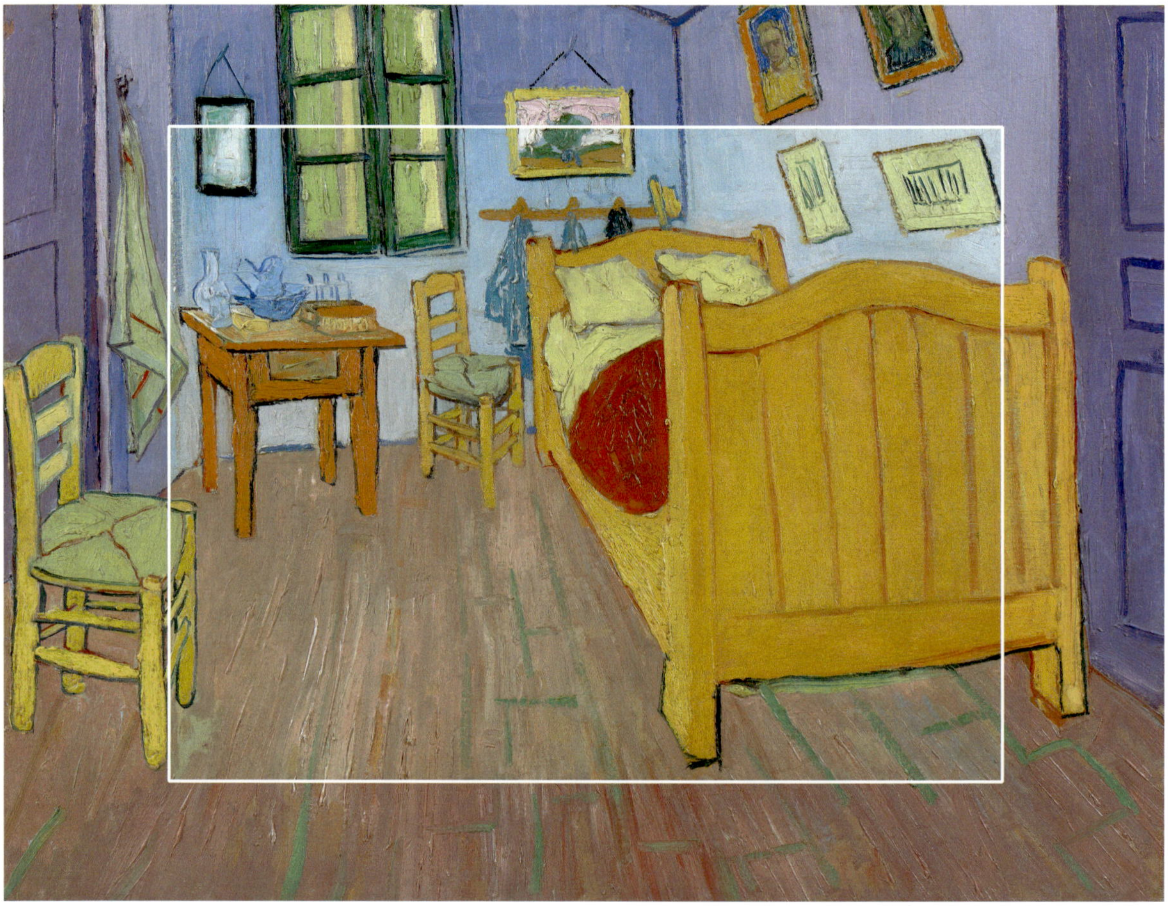

FIGURE 5.13. Vincent van Gogh (Dutch, 1853–1890), *Bedroom,* 1888. Oil on canvas, 72.4 × 91.4 cm (28½ × 36 in.). Amsterdam, Van Gogh Museum. A visualization of how it might have looked when first painted is shown in the outer frame; the center is its current condition.

sold art supplies in Paris, Van Gogh wrote: "You were right to tell Tasset that the geranium lake should be included after all, he sent it, I've just checked—*all the colours that Impressionism has made fashionable are unstable,* all the more reason boldly to use them too raw, time will only soften them too much."[4]

Spectrophotometric measurements of the Van Gogh Museum's *Bedroom* and other paintings by Van Gogh, together with historic cochineal and eosin lake dispersions with linseed oil, made it possible to estimate the absorption and scattering properties of both the lakes and of other

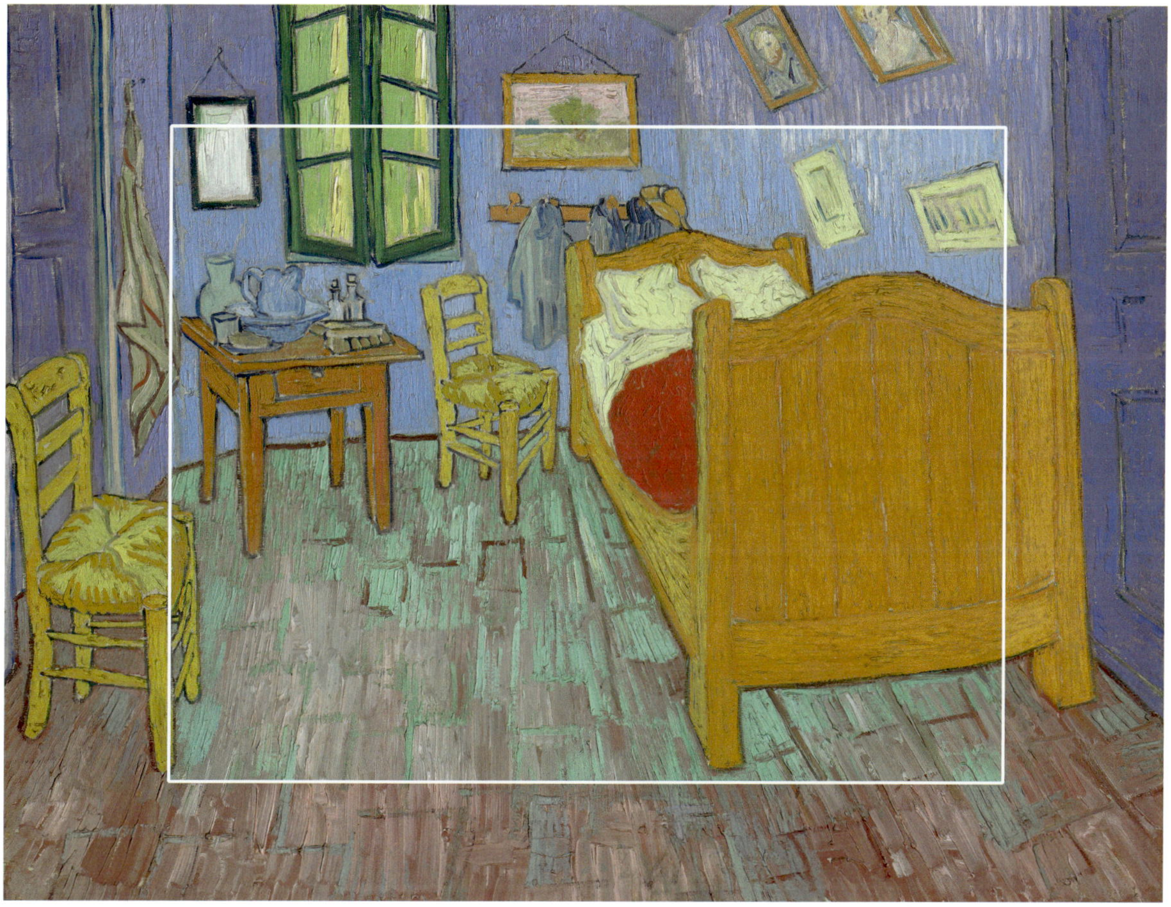

FIGURE 5.14. Vincent van Gogh (Dutch, 1853–1890), *Bedroom*, 1889. Oil on canvas, 73.7 × 93 cm (29 × 36⅝ in.). Art Institute of Chicago, 1926.417. A visualization of how it might have looked when first painted is shown in the outer frame; the center is its current condition.

paints used in the two versions of *Bedroom*. These data were used to create figures 5.13 and 5.14, digital reconstructions of both paintings where passages containing these lake pigments were "unfaded," thus visualizing how these paintings might have looked initially.[5] As the original concentration of the lake pigments is unknown, the estimated concentration was subjective, based on the expertise of painting conservators, curators, and conservation scientists at each museum. These visualizations, albeit speculative, clearly demonstrate the adverse effects of light-induced damage.

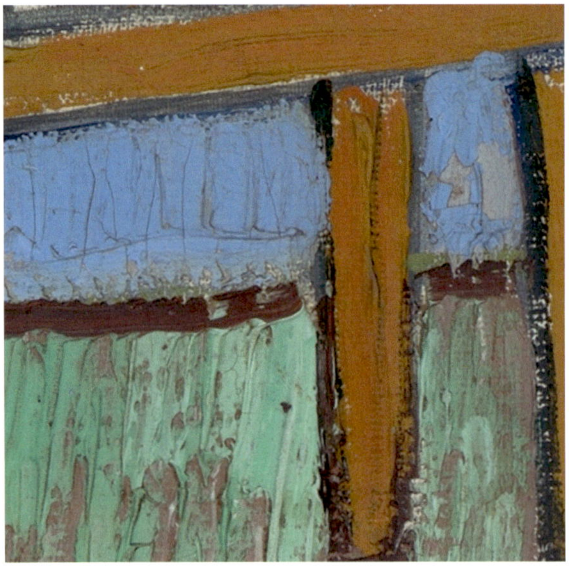

A

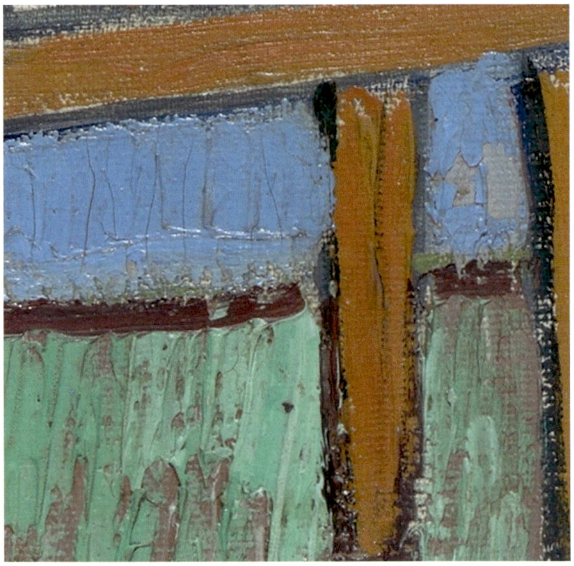

B

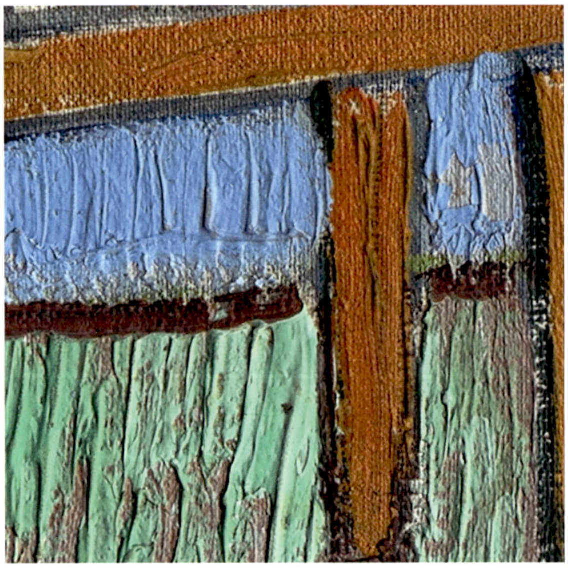

C

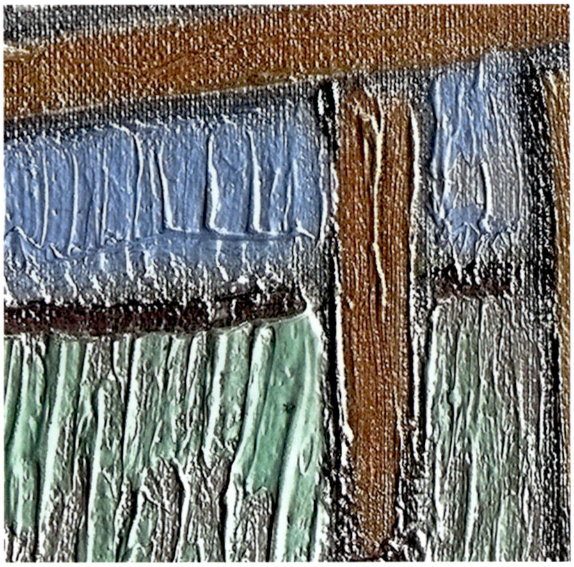

D

FIGURE 5.15. Detail of Vincent van Gogh (Dutch, 1853–1890), *Bedroom,* 1889. Oil on canvas, 73.7 × 93 cm (29 × 36⅝ in.). Art Institute of Chicago, 1926.417. The detail is rendered with (A) diffuse lighting; (B) typical photographic studio lighting; (C) raking collimated light, assuming that the painting is matte; and (D) raking collimated light, assuming that the painting is highly glossy.

Lighting Geometry

The physical distribution of light within a space—in other words, a light source's geometry—also has a profound effect on an artwork's appearance. Lighting has two limiting geometries. At one limit, light is completely diffuse; that is, it is produced equally in all directions. An example is an extremely cloudy day when we would not see shadows. At the other limit, light is collimated where the light is focused in a single direction, producing a narrow beam of light. On an extremely clear day the sun produces highly collimated light, resulting in sharply focused shadows.

Lighting in a museum can be predominantly diffuse, if only indirect light from the ceiling is used, or predominantly collimated, if spotlights are used with minimal ambient illumination. When the walls, ceilings, and floors are light in color, the ambient illumination adds more diffuse light to the direct lighting on the painting than when these building elements are dark. Diffuse lighting masks while collimated lighting accentuates gloss, texture, and impasto, particularly as the lighting moves toward raking angles, as shown in figure 5.15 for a detail of Van Gogh's *Bedroom*.[6]

Viewing Comfort

The color, color rendering, and illuminance of direct and ambient lighting affect the color appearance of artwork. There is a further consideration: viewing comfort. As illuminance and CCT decrease, chromatic adaptation becomes increasingly incomplete and we are more aware of the color of illumination. Although we discount the illumination color when judging objects, there can be a psychological response to the lighting color that affects the viewing experience. We have all had the experience of being outdoors at sunrise and sunset, when CCT and illuminance are low, and in the middle of a sunny day, when CCT is moderate and illuminance is very high. On a cloudy day, CCT is high and illuminance is low. In some geographic locations, winter weather can result in high CCT and low illuminance during much of the day. The existence of seasonal affective disorder, a type of depression linked to lack of exposure to light (Rosenthal et al. 1984), suggests that CCT, illuminance, and mood are interrelated. At museum illuminance levels of 150 lx and less, we are aware of the lighting's color. Research by Philips Lighting during World War II resulted in the designation of a preferred range of CCTs as a function of illuminance; for example, at 150 lx, CCT should range between 2500 and 3300 K (Kruithof 1941). Judgments of "pleasantness" support the preference for low CCTs at low illuminance (Viénot, Duranda, and Mahlerb 2009). Research has identified four underlying dimensions of the effect of lighting on mood: coziness, liveliness, tenseness, and detachment (Vogels 2008; Liu, Luo, and Li 2015). Under typical museum conditions of low CCT and low illuminance, directional lighting resulted in more coziness and liveliness, less tenseness, and equal detachment compared with diffuse lighting. Viewing a traveling exhibition in multiple locations can offer clear evidence that mood and comfort are influenced by physical space and lighting.

Summary

The appearance of artwork in a gallery results from a complex interplay of illuminance, correlated color temperature (CCT), color rendering, lighting geometry, reflectance of building materials, and viewing comfort. Specific choices depend on the physical space, potential for light-induced damage, energy-efficiency considerations, and subjective preferences of lighting designers, curatorial staff, and administrators.

We are in the midst of major changes in lighting as incandescent sources are being phased out because of their low energy efficiency and as solid-state lighting continues to improve. These developments will have a profound effect on museum lighting by expanding choices in CCT while minimizing light-induced damage. Before long it may be possible to design lighting for specific works of art or painting styles while maintaining a consistent CCT within a gallery.

Notes

1. Specifically, 1 fc = 10.76 lx. Lux is defined as units of lumens per square meter, whereas foot-candles are units of lumens per square foot. Foot-candles are no longer used because the International System of Units does not recognize them. Luminance can be converted to illuminance by measuring the luminance of a diffuse white at the location of interest and multiplying by π.
2. In order for lighting to appear neutral white, it must have sufficient luminance and specific colorimetry so that chromatic adaptation is complete (Rea and Freyssinier 2013). For example, the colorimetry of fluorescent lights sometimes changes over time. Suppose a space is lit entirely with fluorescent lights and the lighting appears neutral white. In this case, it is easy to identify a single tube that ages at a fast rate: it appears colored. However, if all the lights age at the same rate, their hue would be less obvious, if seen at all.

3. V. van Gogh, N. Bakker, L. Jansen, and H. Luijten, *The Real Van Gogh: The Artist and His Letters,* Royal Academy of Arts, London. http://vangoghletters.org, number 705.
4. V. van Gogh, N. Bakker, L. Jansen, and H. Luijten, *The Real Van Gogh: The Artist and His Letters,* Royal Academy of Arts, London. http://vangoghletters.org, number 595. The emphasis is Van Gogh's.
5. The approach was similar to the one used to replace the darkened zinc yellow in Georges Seurat's *A Sunday on La Grande Jatte—1884* (Berns, Imai, et al. 2005; Berns, Byrns, et al. 2006).
6. Generating these images required separate image data of the painting's color and topography, obtained using four sequential images of the painting, each having a different lighting direction (Berns and Chen 2012; Cox and Berns 2015).

References

ASTM. 2012. *Standard Terminology of Appearance (E284-12).* West Conshohocken, PA: ASTM International.

Berg, K. J. van den, A. Burnstock, L. Carlyle, M. Clarke, and E. Hendriks. 2006. "Fading of Red Lake Paints after Vincent van Gogh: An Interdisciplinary Study Involving Three De Mayerne Projects." In *Reporting Highlights of the De Mayerne Programme,* edited by J. J. Boon and E. S. B. Ferreira, 89–96. The Hague: Netherlands Organisation for Scientific Research.

Berns, R. S. 2011. "Designing White LED Lighting for the Display of Art: A Feasibility Study." *Color Research and Application* 36: 324–34.

Berns, R. S., S. Byrns, F. Casadio, I. Fiedler, C. Gallagher, F. H. Imai, A. Newman, and L. A. Taplin. 2006. "Rejuvenating the Color Palette of Georges Seurat's 'A Sunday on La Grande Jatte, 1884': A Simulation." *Color Research and Application* 31: 278–93.

Berns, R. S., and T. Chen. 2012. "Practical Total Appearance Imaging of Paintings." In *Archiving 2012,* 162–67. Copenhagen: Imaging Science and Technology.

Berns, R. S., F. H. Imai, L. A. Taplin, S. Byrns, F. Casadio, I. Fiedler, C. Gallagher, and A. Newman. 2005. "Rejuvenating the Appearance of Seurat's 'A Sunday on La Grande Jatte—1884' Using Colour and Imaging Science Techniques: A Simulation." In *14th Triennial Meeting, The Hague, 12–16 September 2005: Preprints (ICOM Committee for Conservation),* edited by I. Verger, 452–58. London: Earthscan.

Burgt, P. van der, and J. van Kemenade. 2010. "About Color Rendition of Light Sources: The Balance between Simplicity and Accuracy." *Color Research and Application* 35: 85–93.

CIE. 1995. *Method of Measuring and Specifying Colour Rendering Properties of Light Sources (13.3-1995).* Vienna: Commission Internationale de l'Eclairage.

———. 2004. *Control of Damage to Museum Objects by Optical Radiation (157:2004).* Vienna: Commission Internationale de l'Eclairage.

Cox, B. D., and R. S. Berns. 2015. "Imaging Artwork in a Studio Environment for Computer Graphics Rendering." In *Proc. SPIE 9398, Measuring, Modeling, and Reproducing Material Appearance,* 939803. San Francisco: SPIE.

Cuttle, C. 2007. *Light for Art's Sake: Lighting for Artworks and Museum Displays*. 1st ed. Amsterdam: Butterworth-Heinemann.

David, A., P. T. Fini, K. W. Houser, Y. Ohno, M. P. Royer, K. A. G. Smet, M. Wei, and L. Whitehead. 2015. "Development of the IES Method for Evaluating the Color Rendition of Light Sources." *Optics Express* 23: 15888–906.

Davis, W., and Y. Ohno. 2010. "Color Quality Scale." *Optical Engineering* 49: 033602.

Dijk, M. van. 2013. "Van Gogh and the Laws of Colour: An Introduction." In *Van Gogh's Studio Practice,* edited by M. Vellekoop, M. Geldof, E. Hendriks, L. Jansen, and A. de Tagle, 216–25. New Haven, CT: Yale University Press.

DiLaura, D. L., K. W. Houser, R. G. Mistrick, and G. R. Steffy. 2011. *The Lighting Handbook: Reference and Application*. 10th ed. New York: Illuminating Engineering Society of North America.

Druzik, J., and S. Michalski. 2012. "The Lighting of Easel Paintings." In *The Conservation of Easel Paintings,* edited by J. H. Stoner and R. Rushfield, 678–92. New York: Routledge.

Geldof, M., M. de Keijzer, M. van Bommel, K. Pilz, J. Salvant, H. van Keulen, and L. Megens. 2013. "Van Gogh's Geranium Lake." In *Van Gogh's Studio Practice,* edited by M. Vellekoop, M. Geldof, E. Hendriks, L. Jansen, and A. de Tagle, 268–89. New Haven, CT: Yale University Press.

Geldof, M., L. Megens, and J. Salvant. 2013. "Van Gogh's Palette in Arles, Saint-Rémy and Auvers-sur-Oise." In *Van Gogh's Studio Practice,* edited by M. Vellekoop, M. Geldof, E. Hendriks, L. Jansen, and A. de Tagle, 238–55. New Haven, CT: Yale University Press.

ISO. 2005. *Photometry: The CIE System of Physical Photometry (ISO 23539:2005; CIE S 010/E:2004)*. Geneva: International Organization for Standardization.

Judd, D. B. 1967. "A flattery index for artificial illuminants." *Journal Illuminating Engineering Society* 62:593–598.

Khanh, T. Q., P. Bodrogi, Q. T. Vinh, and H. Winkler. 2015. *LED Lighting Technology and Perception*. Weinheim, Ger.: Wiley-VCH.

Kruithof, A. A. 1941. "Tubular Luminescence Lamps for General Illumination." *Philips Technical Review* 6: 65–73.

Liu, A., A. Tuzikas, A. Žukauskas, R. Vaicekauskas, P. Vitta, and M. Shur. 2013. "Cultural Preferences to Color Quality of Illumination of Different Artwork Objects Revealed by a Color Rendition Engine." *IEEE Photonics* 5: 6801010.

Liu, X. Y., M. R. Luo, and H. Li. 2015. "A Study of Atmosphere Perceptions in a Living Room." *Lighting Research and Technology* 45: 581–94.

Michalski, S. 1992. "Sharing Responsibility for Conservation Decisions." In *Durability and Change: The Science, Responsibility, and Cost of Sustaining Cultural Heritage,* edited by W. E. Krumbein, P. Brimblecombe, D. E. Cosgrove, and S. Staniforth, 241–58. New York: Wiley.

Miller, N. J., and S. M. Rosenfeld. 2012. *Demonstration of LED Retrofit Lamps at the Smithsonian American Art Museum, Washington, DC: Final Report Prepared in Support of the U.S. DOE Solid-State Lighting Technology Demonstration GATEWAY Program*. Richland, WA: Pacific Northwest National Laboratory.

Nascimento, S. M. C., and O. Masuda. 2014. "Best Lighting for Visual Appreciation of Artistic Paintings – Experiments with Real Paintings and Real Illumination." *Journal of the Optical Society of America A* 31: A214–A219.

Nickerson, D. 1960. "Light Sources and Color Rendering." *Journal of the Optical Society of America* 50: 57–69.

Planck, M. 1901. "Über das Gesetz der Energieverteilung im Normalspectrum" (On the Law of Distribution of Energy in the Normal Spectrum). *Annalen der Physik* 4: 553–63.

Rea, M. S., and J. P. Freyssinier. 2008. "Color Rendering: A Tale of Two Metrics." *Color Research and Application* 33: 192–202.

——. 2010. "Color Rendering: Beyond Pride and Prejudice." *Color Research and Application* 35: 401–9.

——. 2013. "White Lighting." *Color Research and Application* 38: 82–92.

Rosenthal, N. E., D. A. Sack, J. C. Gillin, A. J. Lewy, F. K. Goodwin, O. Davenport, P. S. Mueller, D. A. Newsome, and T. A. Wehr. 1984. "Seasonal Affective Disorder: A Description of the Syndrome and Preliminary Findings with Light Therapy." *Archives of General Psychiatry* 41: 72–80.

Viénot, F., G. Coron, and B. Lavédrine. 2011. "LEDs as a Tool to Enhance Faded Colours of Museums Artefacts." *Journal of Cultural Heritage* 12: 431–40.

Viénot, F., M.-L. Duranda, and E. Mahlerb. 2009. "Kruithof's Rule Revisited Using LED Illumination." *Journal of Modern Optics* 56: 1433–46.

Vogels, I. 2008. "Atmosphere Metrics: Development of a Tool to Quantify Experienced Atmosphere." In *Probing Experience: From Assessment of User Emotions and Behaviour to Development of Products,* edited by J. M. D. M. Westerink, M. Ouwerkerk, T. J. M. Overbeek, W. F. Pasveer, and B. de Ruyter, 25–41. Dordrecht: Springer.

Whitmore, P. M., X. Pan, and C. Bailie. 1999. "Predicting the Fading of Objects: Identification of Fugitive Colorants through Direct Nondestructive Lightfastness Measurements." *Journal of the American Institute for Conservation* 38: 395–409.

Žukauskas, A., R. Vaicekauskas, P. Vitta, A. Tuzikas, A. Petrulis, and M. Shur. 2012. "Color Rendition Engine." *Optics Express* 20: 5356–67.

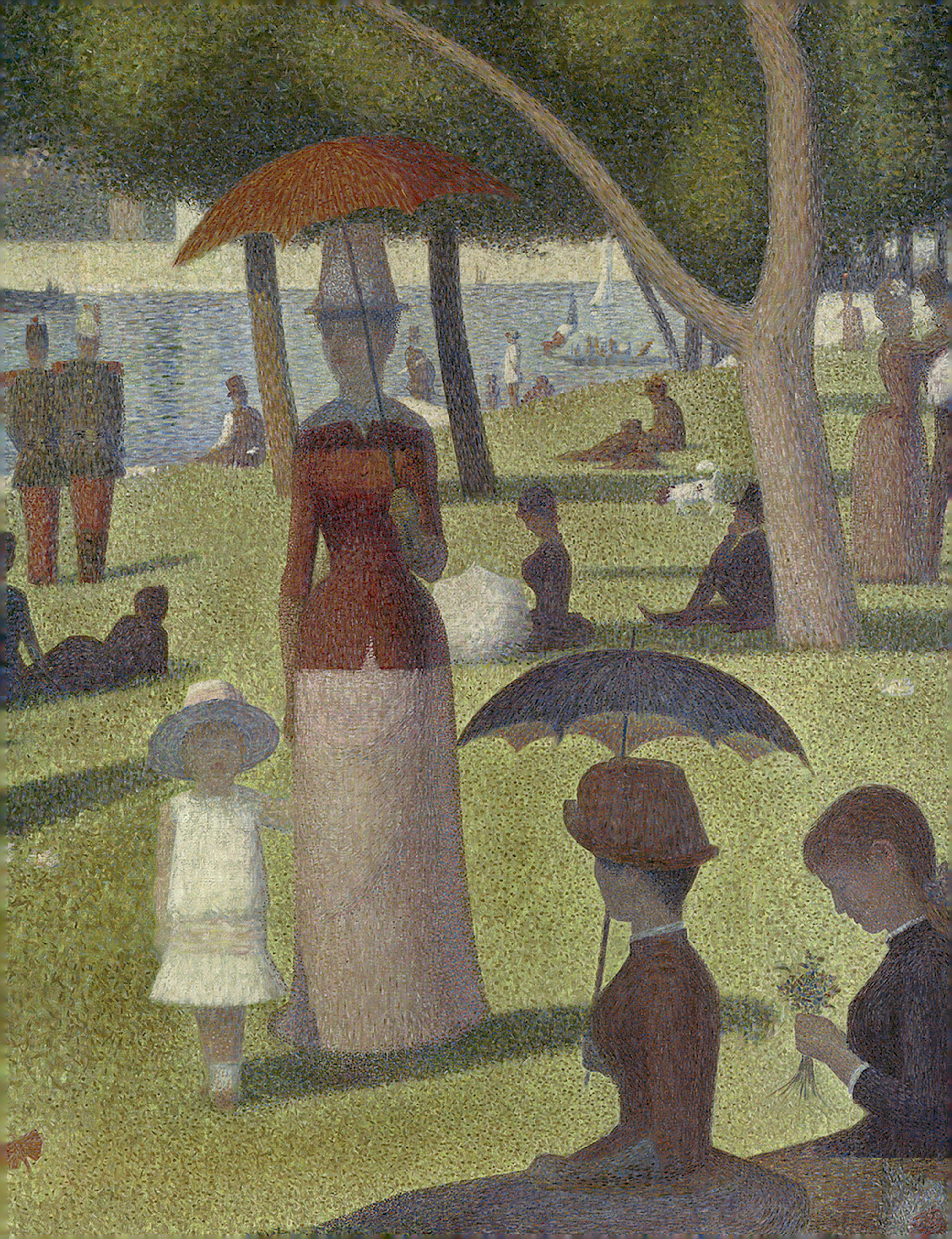

PAINTING MATERIALS

6

Paintings typically consist of a wooden or fabric support, a ground layer, paint layers, and in some cases a varnish layer. When traditional painting materials are used, the final color depends on each paint layer's internal absorption and scattering characteristics as a function of wavelength and on the painting's surface roughness. Internal scattering determines the opacity of a paint layer. Depending on opacity, an artist might use any of various painting techniques, ranging from the application of a single opaque layer to the building up of multiple translucent or transparent layers on top of an opaque ground. There are also paints available that produce colors by fluorescence and interference, and paints with metallic, iridescent, and pearlescent effects.

Selection of the type and number of paints depends on the artists, their ability to control color changes through mixing, and the optical properties of the specific materials. The final color also depends on the presence of spatial patterns that cannot be resolved by our visual system.

Applying a picture varnish will smooth the surface. As a result, colors appear darker and deeper with an overall enlargement in color gamut. When natural varnishes are used, they yellow over time, leading to dramatic color changes.

Although Georges Seurat's *A Sunday on La Grande Jatte—1884* is considered a pointillist icon, the artist used both impressionist and pointillist brushstrokes, readily seen when comparing the little girl's dress with the sunlit grass (detail, fig. 6.16).

Absorption

Traditional colorants produce color by absorbing and scattering incident light selectively across the visible spectrum.[1] A variety of materials are used in paint, including traditional organic and inorganic colorants, fluorescent colorants, ground metals, and interference and pearlescent colorants. Changes in hue are associated, in general, with changes in absorption, while changes in vividness and depth are affected by changes in scattering along with absorption.

We were all taught at an early age that blue and yellow make green. Mixing Prussian blue with cadmium yellow provides clear evidence, producing a vivid green. But if we replace Prussian blue with ultramarine in the mixture, the result is a greenish gray. Each of these blue colorants mixed with white produces a vivid blue. Why, then, is ultramarine such a poor choice if we want to make green?

To create figure 6.1, which explores that question, tint ladders of ultramarine blue and titanium white and of arylide (hansa) yellow and titanium white were used to characterize each paint's absorption and scattering properties as a function of wavelength (Berns and Mohammadi 2007).[2] These properties enable calculating spectral and colorimetric data from concentration. Shifting ultramarine's spectral properties to shorter and longer wavelengths simulated a reddish blue and a greenish blue, respectively. Each blue was mixed with white to achieve a vivid color. That vivid blue and the yellow paint directly from the tube were mixed at various ratios from 100% vivid blue to 100% yellow. The yellow absorbs short wavelengths; the blues absorb long wavelengths. In the middle wavelengths, the amount of absorption depends on the particular blue. Notice that each set of spectra has a single intersecting wavelength within the middle wavelengths. As this wavelength increases in reflectance, the resulting green colors increase in their vividness.[3] Prussian blue, manganese, and phthalocyanine all produce vivid greens because their absorption and scattering properties lead to a high reflectance at the intersecting wavelength. Ultramarine and cobalt blue result in low reflectance, producing dull greens. The most vivid greens are produced using blues that absorb light only at long wavelengths and yellows that absorb light only at short wavelengths. The middle wavelengths are reflected, leading to the perception of green. This analysis could also be performed with similar results using a greenish blue and greenish-to-reddish yellows. Colorants' spectral properties, not their color names, determine color-mixing behavior.

FIGURE 6.1. Reflectance spectra and color gamuts of arylide yellow mixed with a vivid tint of (A) ultramarine, (B) a reddish blue resulting from shifting ultramarine's absorption and scattering properties to shorter wavelengths, and (C) a greenish blue resulting from shifting ultramarine to longer wavelengths. The color gamuts were calculated by intermixing yellow, blue, and titanium white for the corresponding blue.

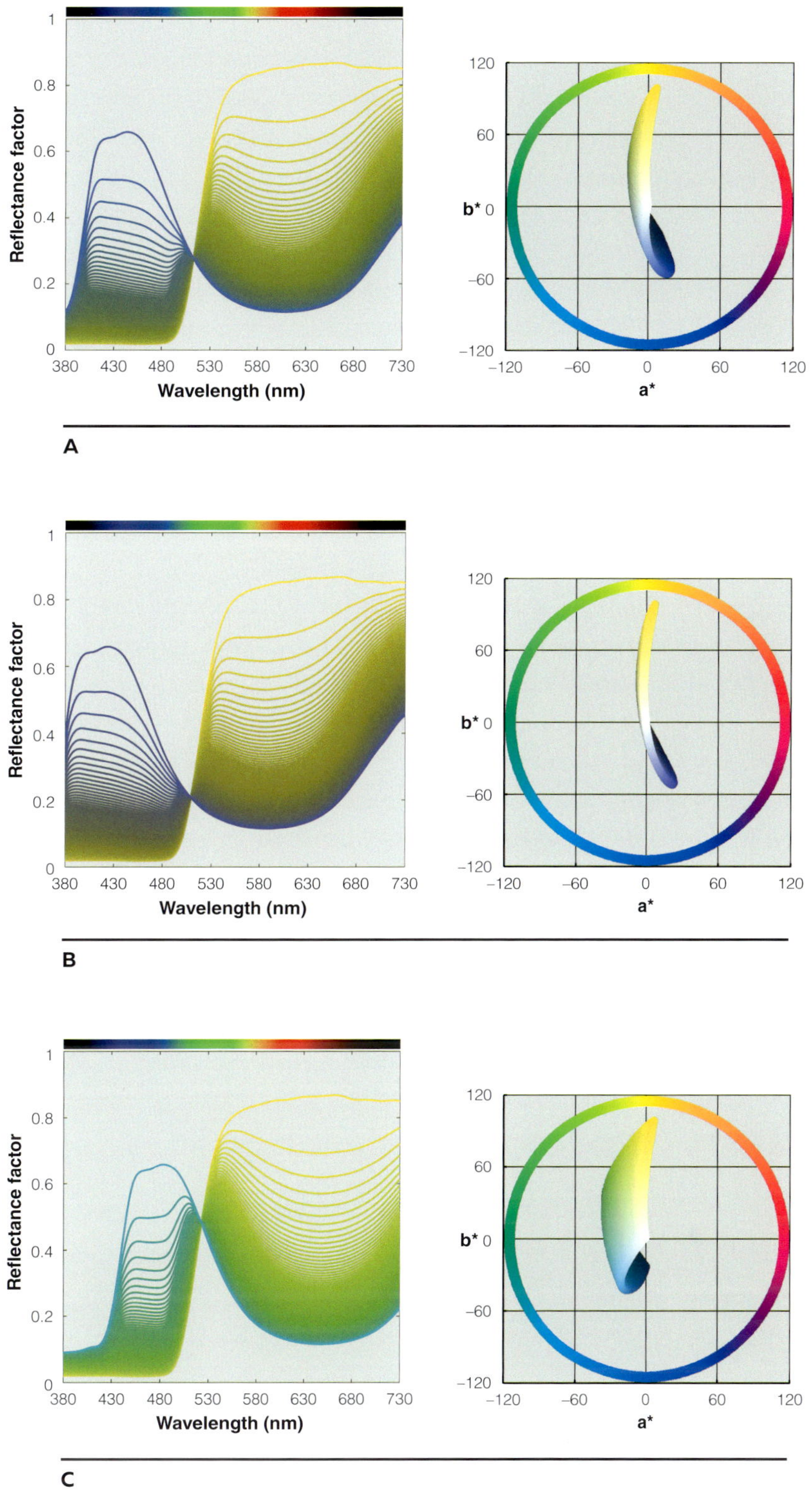

Scattering

Scattering has equally dramatic effects when compared with absorption. In paintings, scattering occurs in two ways. First, light that penetrates into paint can be scattered by its colorants. Applying two different yellow paints at increasing thickness on top of two phthalocyanine blue stripes is shown in figure 6.2. The two yellow paints have identical absorption but different scattering properties. In the top row, the yellow is nearly transparent, similar to a glaze; with increasing thickness, the color changes from cyan to green. In the bottom row, the yellow is nearly opaque, with appreciable scattering. At first the cyan becomes greenish, but as the yellow paint becomes thicker and scatters more light, the cyan becomes yellower. The transparent yellow results in a deeper color than the opaque yellow, as seen when the yellow is applied to the canvas. During the Renaissance, painters often relied on transparent glazes to achieve deeper colors, using, for example, a madder red lake on top of vermilion. Economics also played a role, as in the application of very expensive ultramarine on top of azurite.

The second cause of scattering is surface roughness. All materials reflect a small amount of incident light.[4] When the surface is smooth, we see this as gloss. As the surface becomes rougher, the light reflects more diffusely. When we compare two surfaces of the same material, one glossy and one matte, the matte region looks as if white paint has been added to it. For figure 6.3, matte black and red were painted as stripes, and a

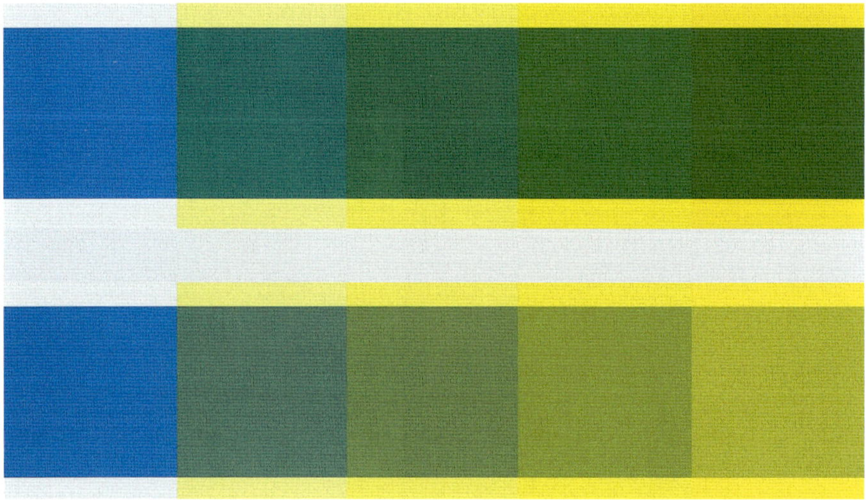

FIGURE 6.2. Transparent (top row) and translucent (bottom row) yellow paints are applied to cyan stripes in increasing thickness from left to right.

high-gloss varnish was applied in the center of each stripe. The varnish reduces the surface roughness, producing a smoother surface (Berns and de la Rie 2003a, 2003b).[5] The light that was reflecting in many directions because of the rough surface now reflects in a single direction, producing a glossy surface. When the stripes are viewed at an angle that keeps the gloss from being seen, as shown in this figure, the black becomes darker and the red becomes deeper.

Scattering within the paint layer and at its surface affects color gamut. Surface roughness has the largest impact, as seen in figure 6.4, where changes in either internal scattering or surface roughness are compared for mixtures of yellow, magenta, black, and white matte paints. Reducing surface roughness by applying a glossy picture varnish enlarged color gamut to a much greater extent than reducing internal scattering. The changes in the lightness of the black paint—minimum L*—are very noticeable.

The surface roughness of a varnish also affects the sharpness of the underlying paint layers. This was explored by evaluating a photograph of a Rembrandt painting seen through glass with differing surface roughnesses, as shown in figure 6.5 (Berns and de la Rie 2003a). Roughness was changed by applying a smoothing or nonsmoothing varnish to sandblasted glass.[6] As the glass became smoother, the painting appeared sharper.

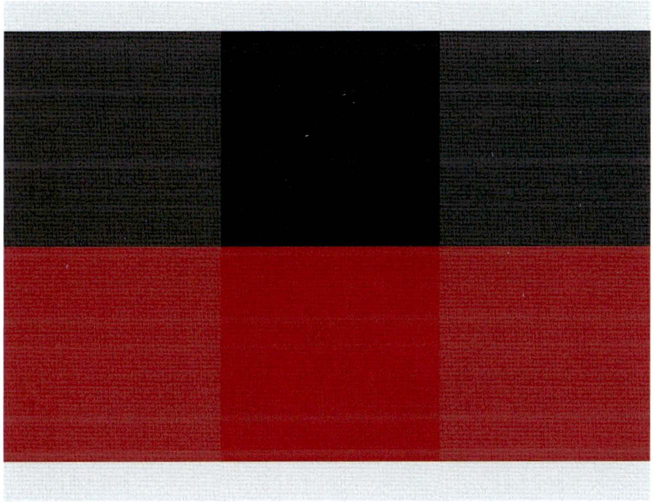

FIGURE 6.3. A glossy varnish is applied to the center of matte black and red stripes of paint.

FIGURE 6.4. Color gamuts of arylide yellow, quinacridone magenta, carbon black, and titanium white matte acrylic dispersion paints, where either (A) the scattering and surface roughness were unchanged, (B) the internal scattering was reduced fivefold and the roughness was unchanged, or (C) the internal scattering was unchanged and the surface roughness was reduced by applying a glossy varnish.

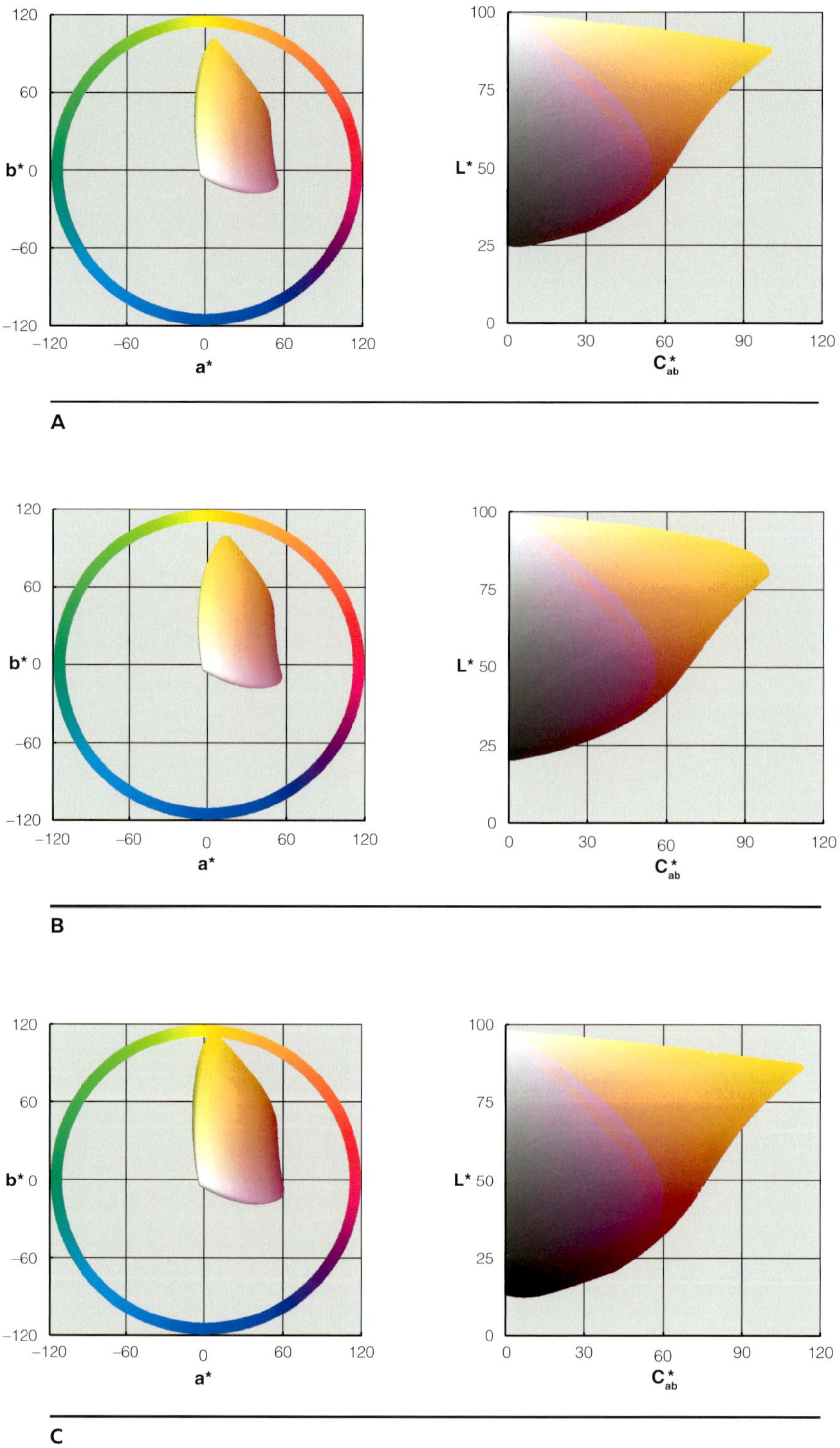

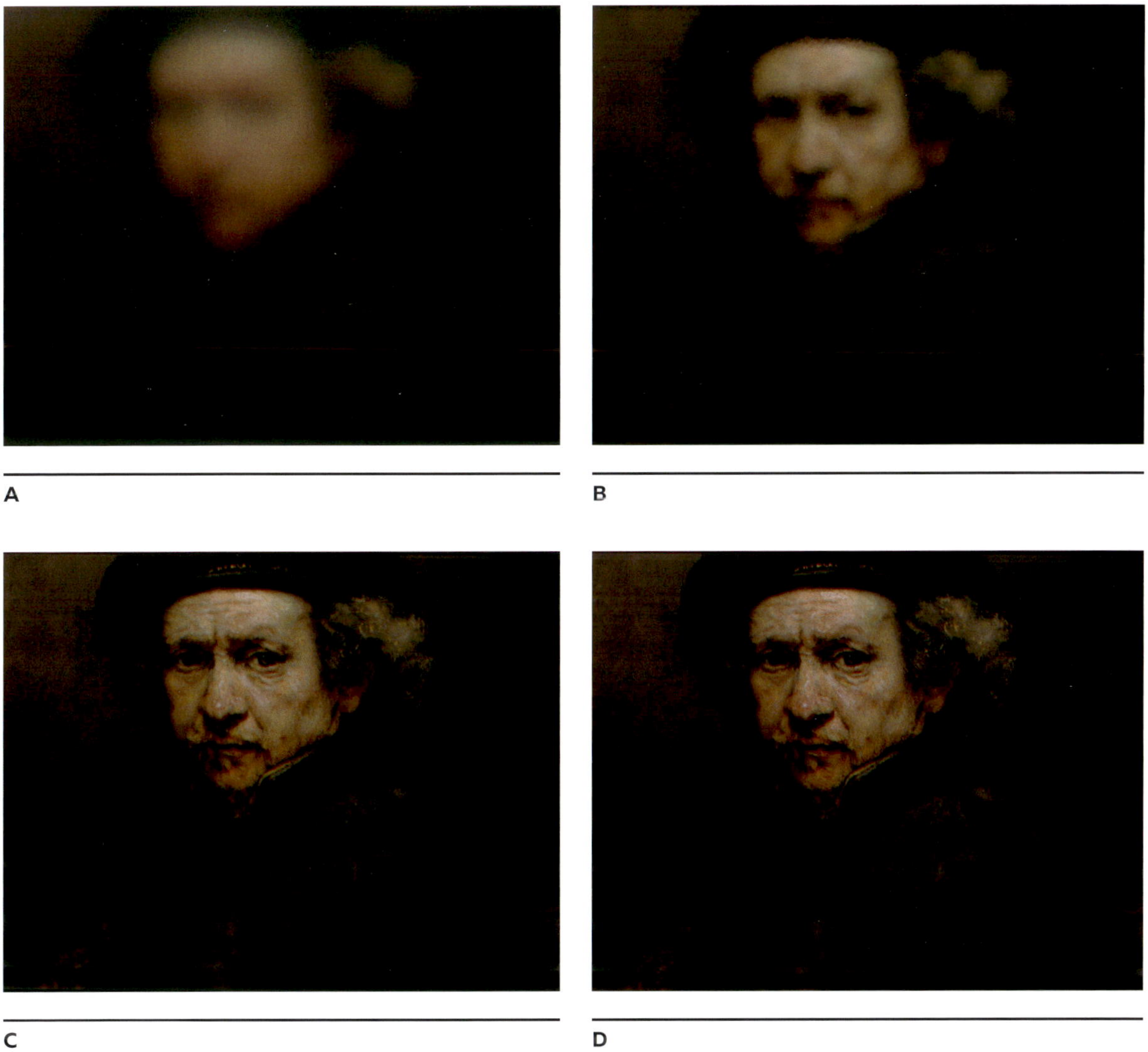

FIGURE 6.5. Detail of Rembrandt van Rijn (Dutch, 1606–1669), *Self-Portrait,* 1659. Oil on canvas, 83.8 × 66 cm (33¼ × 26 in.). Washington, National Gallery of Art, 1937.1.72. For these images, a photograph of the painting is viewed through (A) sandblasted glass, (B) sandblasted glass coated with a nonsmoothing varnish, (C) sandblasted glass coated with a smoothing varnish, or (D) clear glass.

Yellow Colorants and Varnishes

Of all the colors, yellow is the most misunderstood. Most books concerned with color mixing portray yellow as a single wavelength plucked from the visible spectrum rather than a collection of wavelengths ranging from green through red, even though such information has been readily available since the late nineteenth century (Rood 1879).

A popular modern yellow pigment is arylide yellow, developed during the early twentieth century. Its spectrum, shown in figure 6.6, is typical of most yellow objects: light is absorbed in the short wavelengths and reflected in the middle and long wavelengths. It appears yellow because the yellow opponent response is stimulated. It is a reddish yellow because the red opponent response is stimulated slightly.

To create figure 6.7, a tint ladder of arylide yellow and titanium white was produced, and each sample was measured with a spectrophotometer. Titanium white reflects more than 90% of incident light throughout most of the visible spectrum except at short wavelengths, where there is strong absorption.[7] As the yellow concentration increases, the amount of

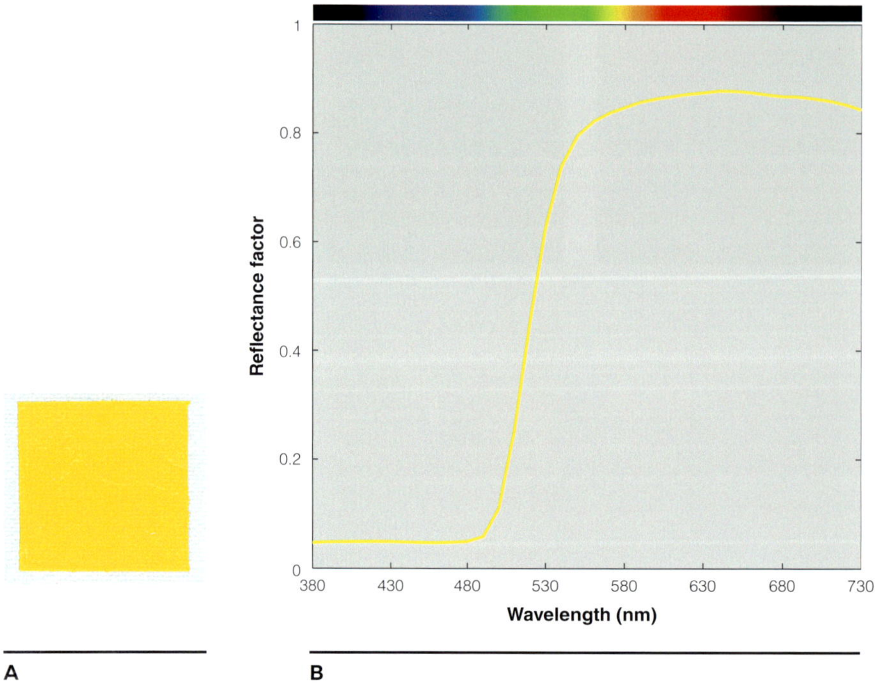

FIGURE 6.6. (A) Arylide yellow painted sample, with (B) its reflectance spectrum.

short-wavelength light that is reflected decreases; that is, the amount of absorption increases. This yellow paint modulates short-wavelength light. It does not appreciatively change the amount of light in the yellow region of the spectrum. Increasing short-wavelength reflectance changes our yellow-opposed-to-blue channel from yellow toward a balance of blue and yellow —that is, neutral, which for this example is white.

Yellow is a psychological primary, and many colorants that appear yellow can be greenish or reddish compared with unique yellow. Reflectance spectra appearing greenish to reddish yellow have a similar shape, absorbing short-wavelength light and reflecting longer-wavelength light with a sharp transition between the two. As this reflectance transition moves to longer wavelengths, the appearance changes from greenish yellow to reddish yellow, shown in figure 6.8.[8] There is also a darkening of the paints. For example, cadmium yellow light is greener and lighter than cadmium yellow deep.

Single wavelengths that appear yellow range from 570 nm to 590 nm (Kuehni 2012). It is possible to produce a reflectance spectrum

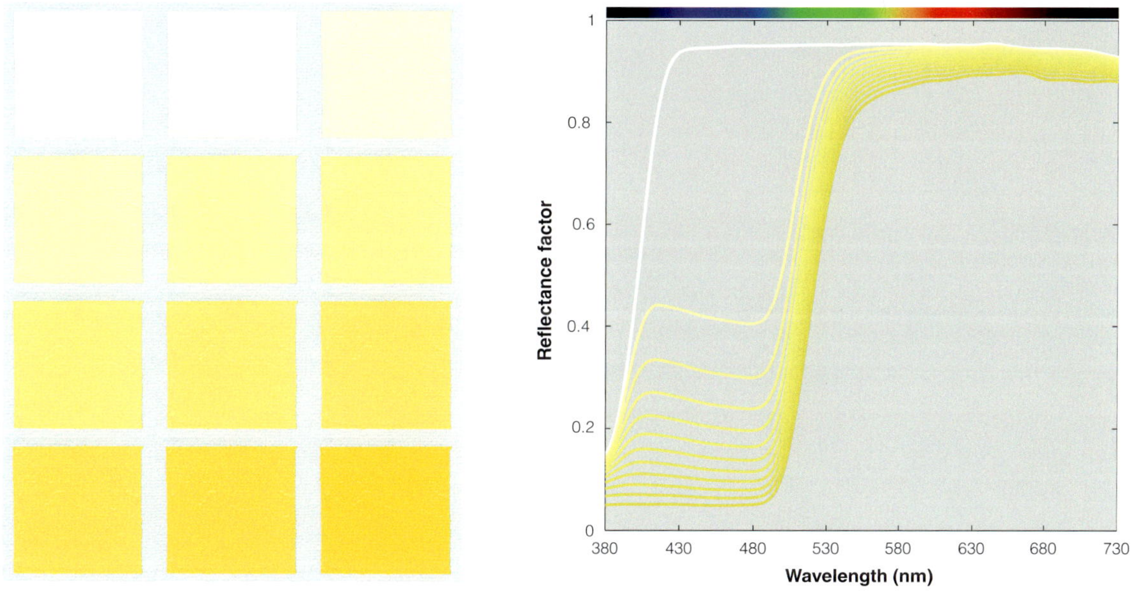

A

B

FIGURE 6.7. (A) Tint ladder of arylide yellow and titanium white, with (B) its reflectance spectra. With an increase in the concentration of yellow, absorption of short-wavelength light increases, reducing reflectance.

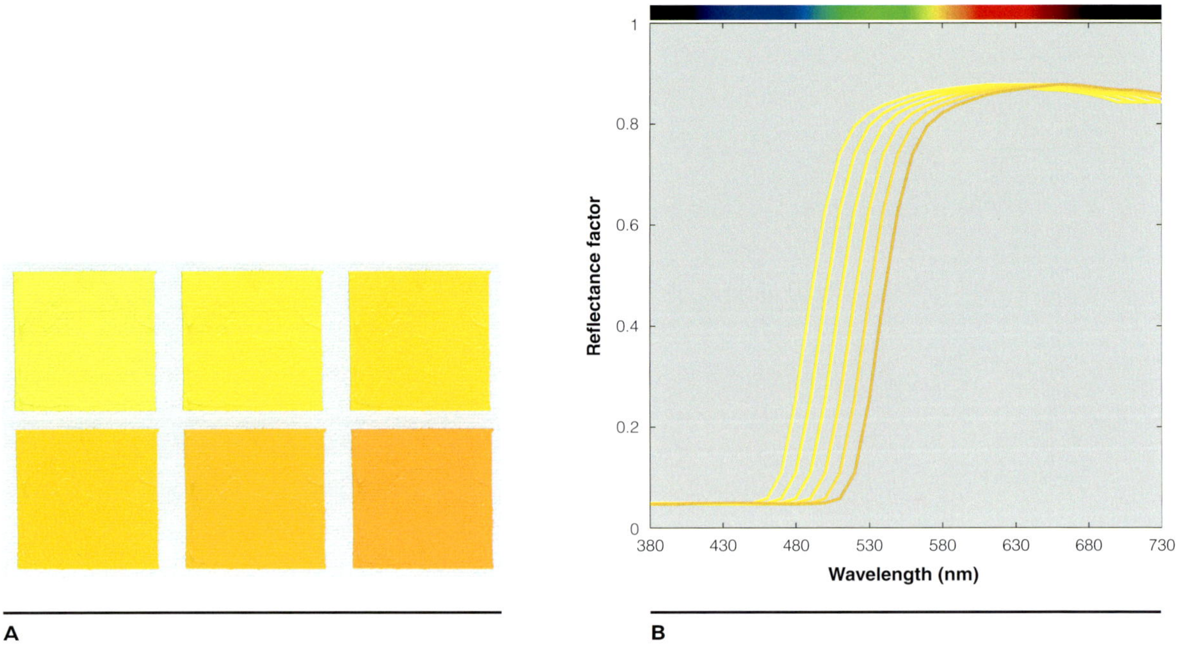

FIGURE 6.8. (A) Yellow paints (from greenish to reddish), with (B) their reflectance spectra.

where these wavelengths are removed. In figure 6.9, a thin piece of didymium glass,[9] a rare-earth mixture used to make fashion and welder safety glasses, is placed on top of yellow paint. The interesting feature of didymium glass is its strong absorption from 560 nm to 600 nm, effectively absorbing all yellow wavelengths. Despite the elimination of all the yellow wavelengths, yellow is clearly seen![10] Certainly spectral yellow exists in nature; however, the vast majority of yellow objects absorb short-wavelength light while reflecting middle- and long-wavelength light. The key is stimulating the yellow opponent response, not producing only yellow wavelengths.

Using yellow colorants is not the only way to have a painting appear yellowish. There is the unintentional yellowing of artwork where a paint medium or varnish yellows over time. Picture varnishes, especially those made from natural resins, will change color, lose gloss, reduce transparency, and fracture at a microscopic level when exposed to light and oxygen (de la Rie 1987, 1988, 1989). The net result is a yellowing, darkening, and loss of sharpness of the underlying paint layers. The best-known example is Leonardo da Vinci's *Mona Lisa* (Mohen, Menu, and Mottin 2006; Elias and Cotte 2008). Exposing a varnish to UV radiation at an

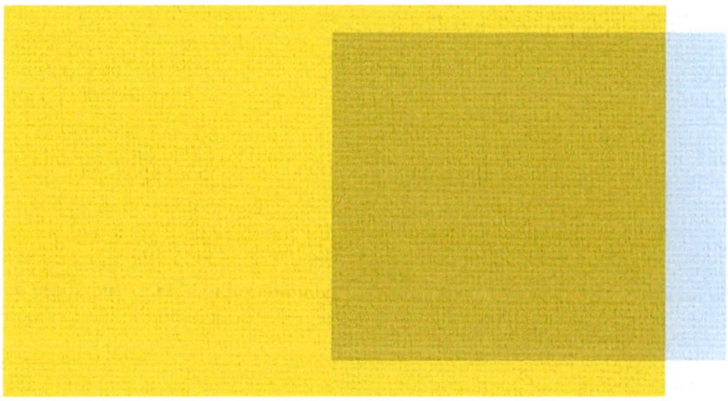

A

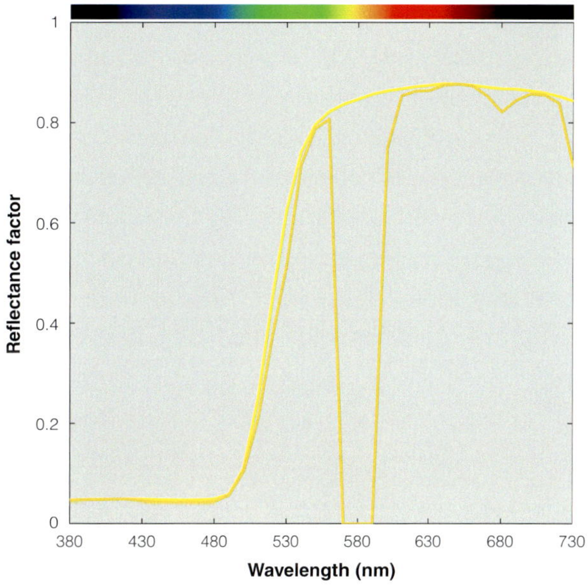

B

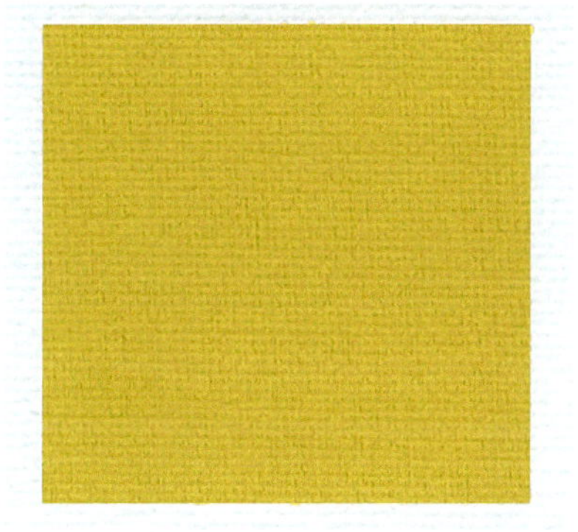

C

FIGURE 6.9. (A) A piece of didymium glass is placed over arylide yellow paint and (B) measured with a spectrophotometer. (C) When the filter perfectly covers the yellow paint, its yellow color is seen clearly. The didymium removes yellow-appearing wavelengths, yet the color is still yellow.

elevated temperature can approximate color changes caused by natural aging (Cotte and Dupraz 2006). With aging, as shown in figure 6.10, the varnish's color changes from colorless to slightly yellowish to deep orange.

An image of Auguste Renoir's *Albert Cahen d'Anvers* was used to simulate the application of an aged varnish in figure 6.11. As varnish ages, the painting becomes browner. The blue passages are nearly black. Although we might compensate for the orange color balance in much the same way that we cognitively discount the color of illumination (as described in chapter 2), we could not return the blue hue to the suit of clothes.

A

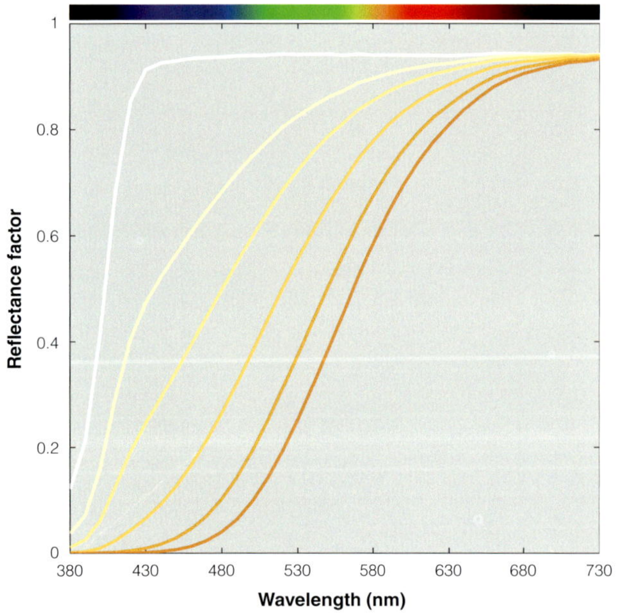

B

FIGURE 6.10. (A) Accelerated-aged natural resin varnish applied on top of white paint. The varnish colors change from colorless to orange over time. (B) A corresponding change in spectral reflectance where absorption increases toward longer wavelengths.

FIGURE 6.11. Pierre-Auguste Renoir (French, 1841–1919), *Albert Cahen d'Anvers*, 1881. Oil on canvas, 79.9 × 63.8 cm (31½ × 25⅛ in.). Los Angeles, J. Paul Getty Museum, 88.PA.133. (A) The effect of applying an accelerated-aged natural varnish on the painting in a rectangular band is simulated. (B) A plot of CIELAB coordinates of the entire painting looking down the L* axis (before varnishing). (C) A plot of CIELAB coordinates after simulating an application of the aged varnish over the entire painting. The aged varnish greatly reduces the painting's color gamut and shifts its color balance toward brown.

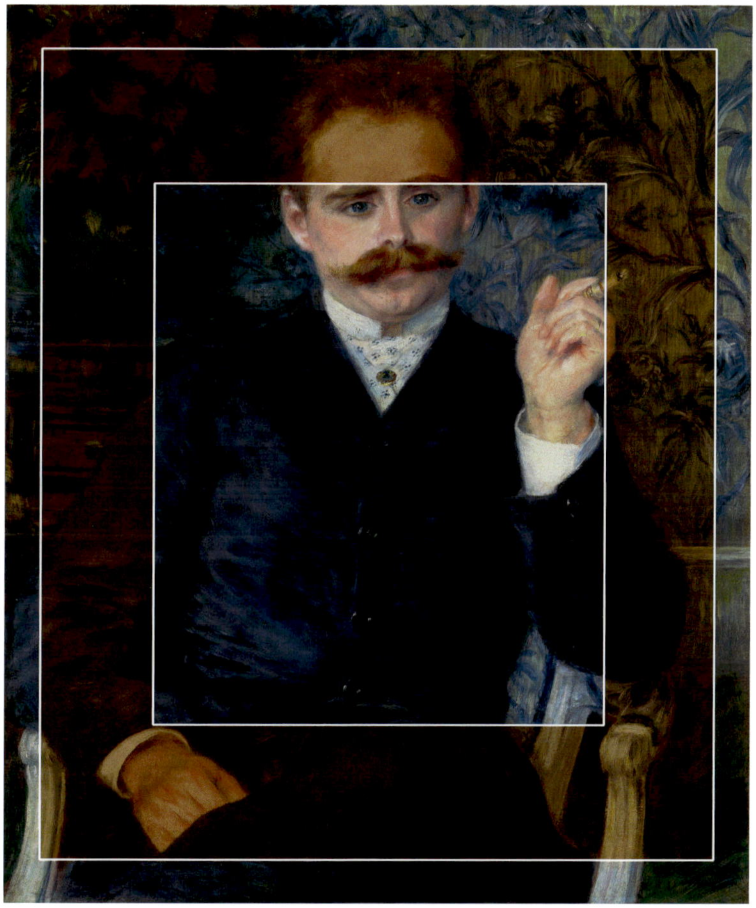

A

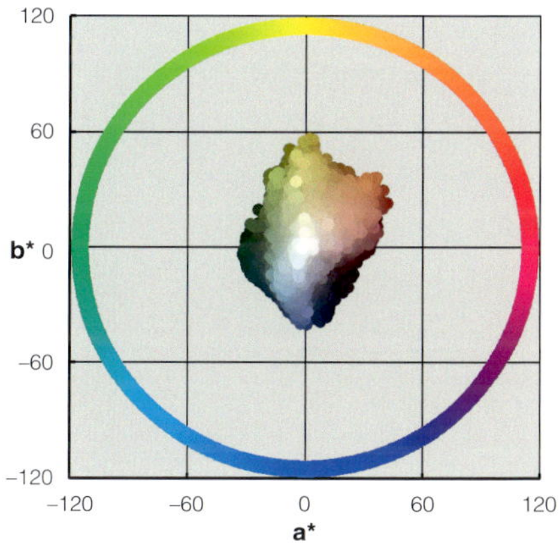

B

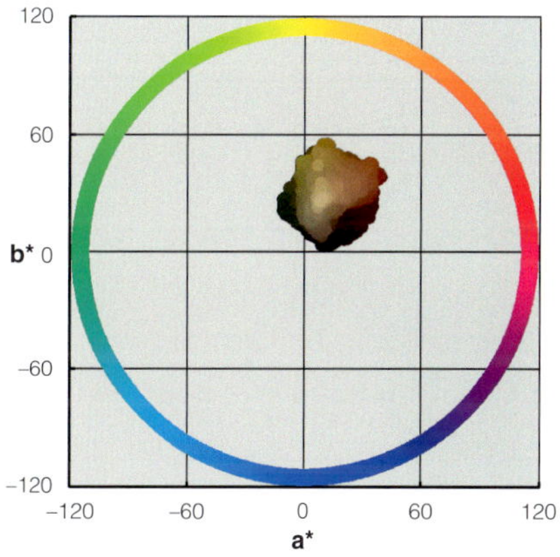

C

Effect Pigments and Fluorescent Colorants

Traditional absorbing and scattering colorants produce color perceptions that are largely independent of geometry when viewed away from gloss angles. Their reflectance spectra are largely invariant with illumination geometry. Today, artists have ready access to materials whose spectra vary with geometry, such as iridescent and interference colorants. Other materials, such as metallic and pearlescent colorants, affect texture within the paint. Colorants whose appearance changes with geometry are called effect pigments.

Since the Egyptians' first use of gold overlays in the decorative arts, metals have been used as artist materials. A distinguishing feature of metals is that their color is best observed at gloss (specular) angles.[11] Adhesion problems with thin-beaten metal overlays caused them to be ground into small particles and dispersed in a medium, producing paint. The mirror-like properties of polished metal were replaced by a diffuse appearance that varied with the size and shape of the metal particles, their surface roughness, and their orientation within the medium.

During the 1980s, aluminum flake, so named because of its similarity to cornflakes when viewed under magnification, could be manufactured with uniform size and high polish; it became a common additive in automotive paint finishes as new application methods were developed that could better control its orientation, leading to more dramatic and consistent changes in appearance. When viewed at near-specular angles,[12] such paints appear much lighter than when they are viewed away from specular; as a result, three-dimensional contours are accentuated. Aluminum flake remains a common additive in automotive finishes. Today there is a range of metallic artist paints containing bronze, copper, zinc, stainless steel, nickel, graphite, and aluminum.

Developments in materials science plus the ability to apply extremely thin inorganic coatings in multiple layers on a variety of substrates resulted in the development of iridescent, pearlescent, and interference-effect pigments (Pfaff 2008). These colorants are used in a variety of applications such as coatings, cosmetics, printing, ceramics, glass, and currency. Changing illumination or viewing geometry best reveals these effects, giving rise to new terms such as color travel, sparkle, glint impression, graininess, and diffuse coarseness (Kirchner et al. 2007; Dekker et al. 2011). The most dramatic changes are achieved using interference colorants where

FIGURE 6.12. Matte paint containing an interference pigment. (Obtaining two colors in a single image was accomplished by wrapping the paint applied to paper around a cylinder and illuminating the cylinder with light through a narrow rectangular aperture.)

hue varies between green and purple, magenta and gold, and blue and red, as shown in figure 6.12. The high cost of interference pigments, combined with their unique color shifts, has led to their usage in currency as a deterrent to counterfeiting. The use of these materials in artworks will be quite challenging for conservators, particularly if imitative inpainting is required. Differences in the orientation and concentration of these pigments affect their final appearance. Matching will be required at different geometries.

Fluorescent colorants are another type of artist material; they absorb light at one wavelength and emit some of it at longer wavelengths, resulting in reflectances greater than 100%. Our visual system recognizes that more light is reflected than is incident, producing a glowing effect. The most common fluorescent colorants are fluorescent whitening agents, or FWAs (Mustalish 2000). These colorants absorb ultraviolet radiation (the excitation region) and emit—fluoresce—in the visible region at short wavelengths, shown in figure 6.13 for "bright white" paper illuminated by skylight. FWAs are used both to counteract the natural yellowing of paper, textiles, and plastics that occurs with time and to increase whiteness.[13] Many modern materials contain FWAs, as in the case of our clothing when it is washed with detergents that "brighten and whiten." Because the fluorescent emission is added to the nonfluorescent reflectance, spectrophotometric measurements are greater than unity.[14]

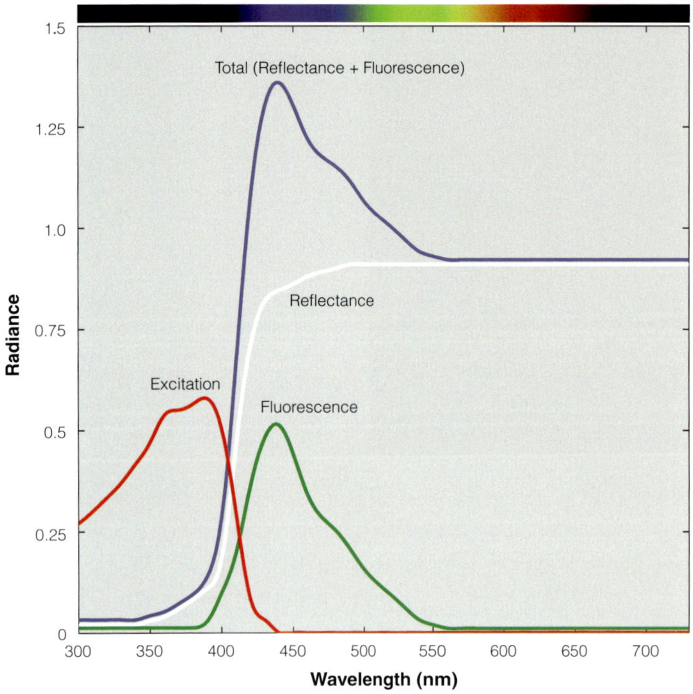

FIGURE 6.13. Spectral properties of "bright white" paper illuminated by skylight. Any absorbed radiation within the excitation region leads to fluorescence. The fluorescence occurs across a range of wavelengths. The reflectance spectrum is the property of the paper without the FWA. The paper's appearance is determined by its total radiance, the fluorescent and nonfluorescent components added together.

As shown in figure 6.14, fluorescent colorants can also be chromatic, like those developed in the 1930s by the Switzer brothers and sold as Day-Glo paints beginning in the 1950s (Barton 2009). Their fluorescence was considerably greater than other colorants known to fluoresce, such as natural Indian yellow (Johnston-Feller 2001). In these cases, excitation occurs in both the UV and visible regions, and the fluorescence occurs in the visible region. Their excitation and emission spectra are similar to those shown in figure 6.13 but are shifted to longer wavelengths, and as in the case of FWAs, their spectra exceed unity.

Because the quantity of light absorbed affects the quantity of light emitted, the fluorescent effect varies with the spectral properties of the lighting. The fluorescent effect of these paints is stronger in daylight

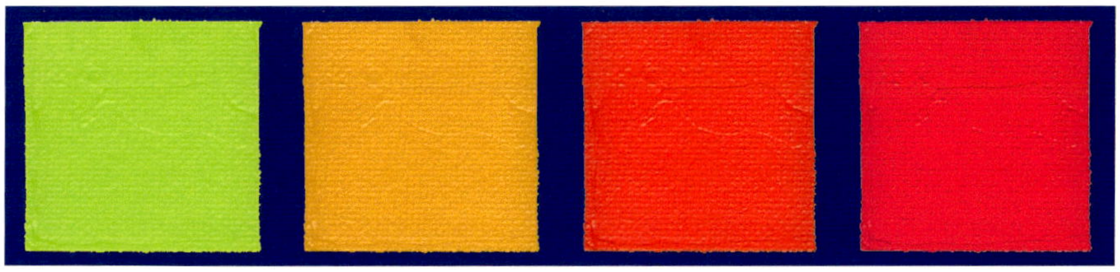

A

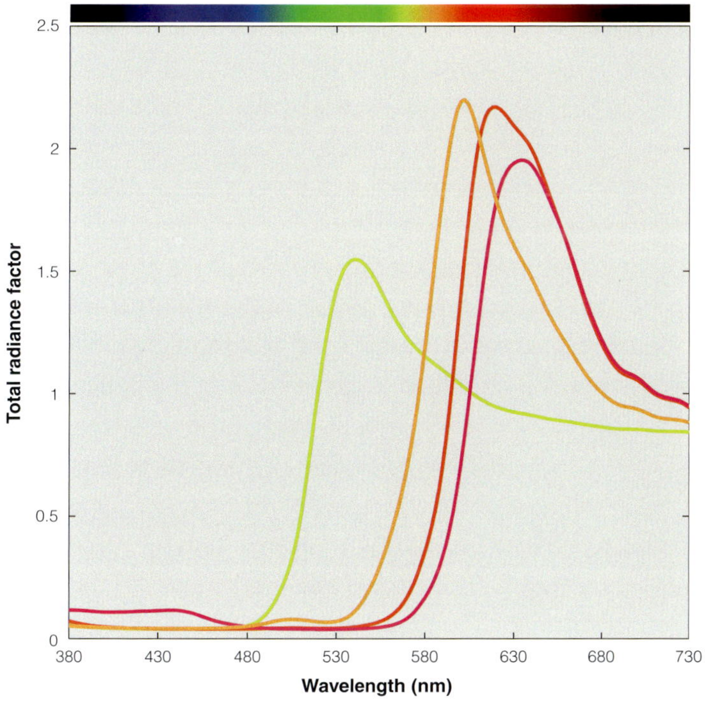

B

FIGURE 6.14. (A) Fluorescent paints illuminated by predominantly UV radiation, with (B) their total radiance measurements when illuminated by skylight.

than under incandescent museum lighting. Although the use of fluorescent colorants in artwork generates visual interest and increased vividness compared with conventional colorants, these colorants when first introduced had poor lightfastness and tended to lose their fluorescent effect over time (Ellis, McGlinchey, and Chao 2002; Morris and Whitmore 2007). Those responsible for lighting older fluorescent artwork face a tradeoff between the immediate visual experience and the longevity of the artwork. Fortunately, the technology has greatly improved, and the lightfastness of modern fluorescent colorants is now similar to that of their nonfluorescent counterparts.

Spatial Color Mixing

In chapter 2 we learned that our visual system processes color spatially and that our three opponent channels have different acuity. The black-opposed-to-white channel has the highest acuity, and the yellow-opposed-to-blue channel has the lowest. We can visualize this as each opponent channel being composed of an area of cone receptors, called a receptive field. The black-opposed-to-white receptive field is the smallest, and the yellow-opposed-to-blue is the largest. Within the receptive field, further discrimination is not possible. Suppose we have two spots of identical light that fall in different locations but still within a receptive field. The visual response to the spots would be the same as to a single spot with twice the amount of light. This means that mixing light in the same spatial location results in the identical color to that of the light being separated in space as long as the stimuli are within a receptive field. There is another factor contributing to spatial mixing: our visual system is not a perfect imaging system; it has a maximum spatial resolution, beyond which blurring occurs.

Spatial mixing is used to great effect in art conservation. When inpainting is done, two approaches are most common (Nadolny 2012). The first is imitative inpainting, where the goal is to match the surrounding area as closely as possible, considering not only the color but also any optical effects that can result from multiple paint layers with dissimilar absorption and scattering. The second approach is discernible inpainting, where, at a distance, the inpainted area appears to match the surround, but close up, spatial patterns including hatching or tiny dots are seen, clearly indicating a conservation treatment. This is an example of spatial mixing. If the difference in lightness between the background color and the individual marks of the spatial pattern is small, then the distance required for spatial mixing will be less than if the lightness difference is large, given the high acuity of our black-opposed-to-white channel. Matching the surrounding area in lightness is more important than matching it in hue and chromatic intensity, again because of the high acuity of our black-opposed-to-white channel. Figure 6.15, Simone Martini's *Saint Luke,* serves to illustrate this point. At the place where the panel joins the frame, alternating red and green stripes, called *tratteggio* (Italian, meaning hatching), spatially mix to produce a golden color that nearly matches the hue of the adjacent gold leaf.[15] Blurring the *tratteggio* and surround, either with image processing or by squinting or taking off reading glasses, can be helpful in determining the quality of the match.

What about the pointillists? Many color books have used Georges Seurat's *A Sunday on La Grande Jatte—1884,* shown in figure 6.16, as the archetype of spatial color mixing. In fact, the painting transitions between Impressionism and Pointillism (Herbert and Harris 2004). *La Grande Jatte* was painted in three stages over a period of five years. In the first stage, the painting was executed using typical impressionist brushwork. In the second stage, Seurat added pointillist brushwork and altered the figures and animals by enlarging or adding contours to their forms. In the third stage, Seurat enlarged the canvas and added a pointillist frame. Notice that the pointillist brushwork is composed of dabs and dashes, not solely of the circular dabs usually associated with this painting style. Although the painting is transitional, at normal viewing distance it appears predominantly pointillist because of its speckled appearance. The painting must be viewed from very far away for the colors to visually merge, demonstrating spatial mixing.

Ironically, unexpected color changes after only a few years increased the painting's pointillist appearance. Seurat called this style "divisionism," because elements are divided into spatial patterns of contrasting colors. One aspect of divisionism was the concept of using paint colors that were quite vivid, simulating colored light. Seurat used red lake, vermilion, chrome yellow, zinc yellow, emerald green, ultramarine, and lead white for *La Grande Jatte's* pointillist brushwork (Fiedler 1984). In the grass, the pointillist dabs contained zinc yellow, applied either directly from the tube or mixed with emerald green. The result, a shimmering effect of colored

FIGURE 6.15.
(A) Simone Martini (Italian, ca. 1284–1344), *Saint Luke*, 1330s. Tempera and gold leaf on panel, 55.9 × 35.6 cm (22¼ × 14½ in.). Los Angeles, J. Paul Getty Museum, 82.PB.72.
(B) Detail of the upper right edge of the panel. Missing paint has been inpainted with alternating red and green stripes, which at a distance produce a color similar to the surrounding gold leaf.
(C) Blurring the image simulates the spatial mixing.

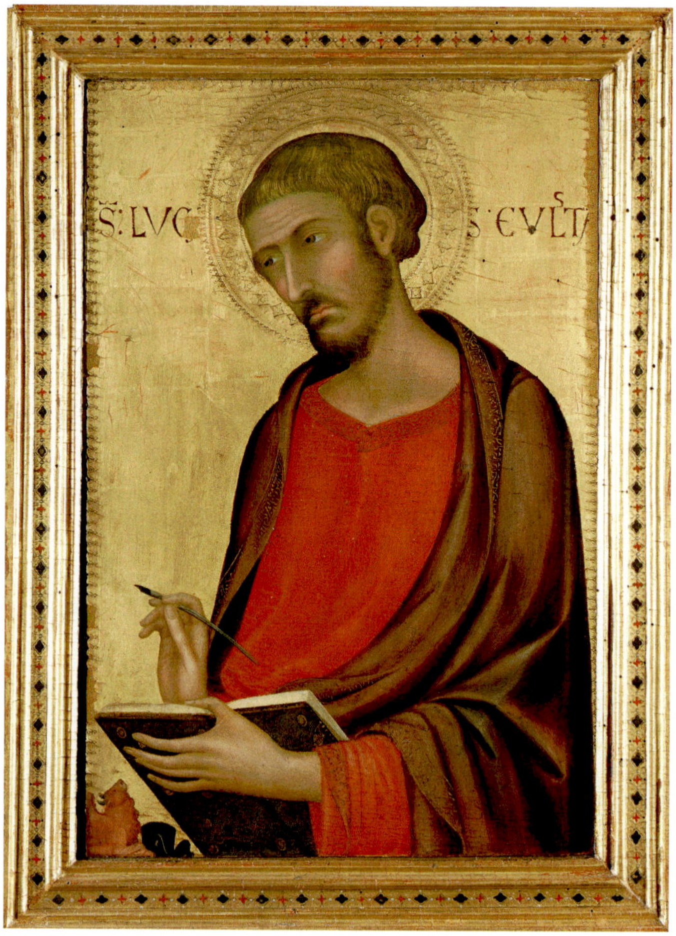

A

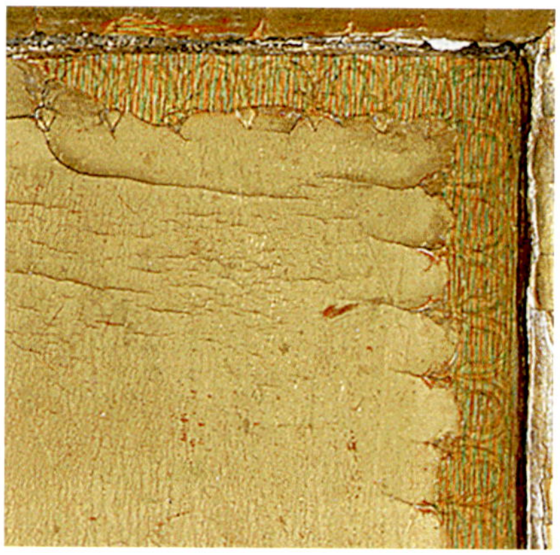

B

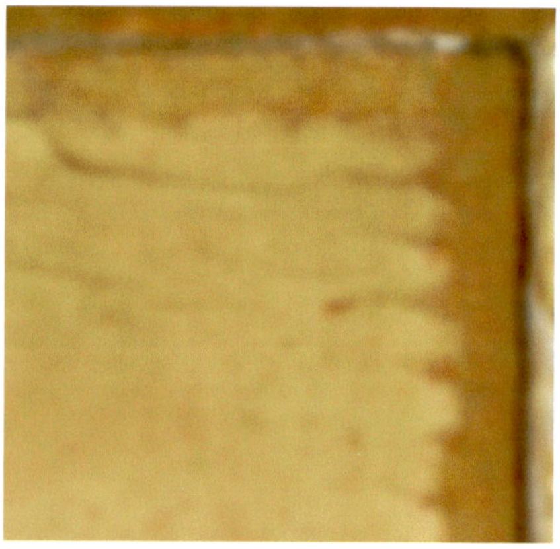

C

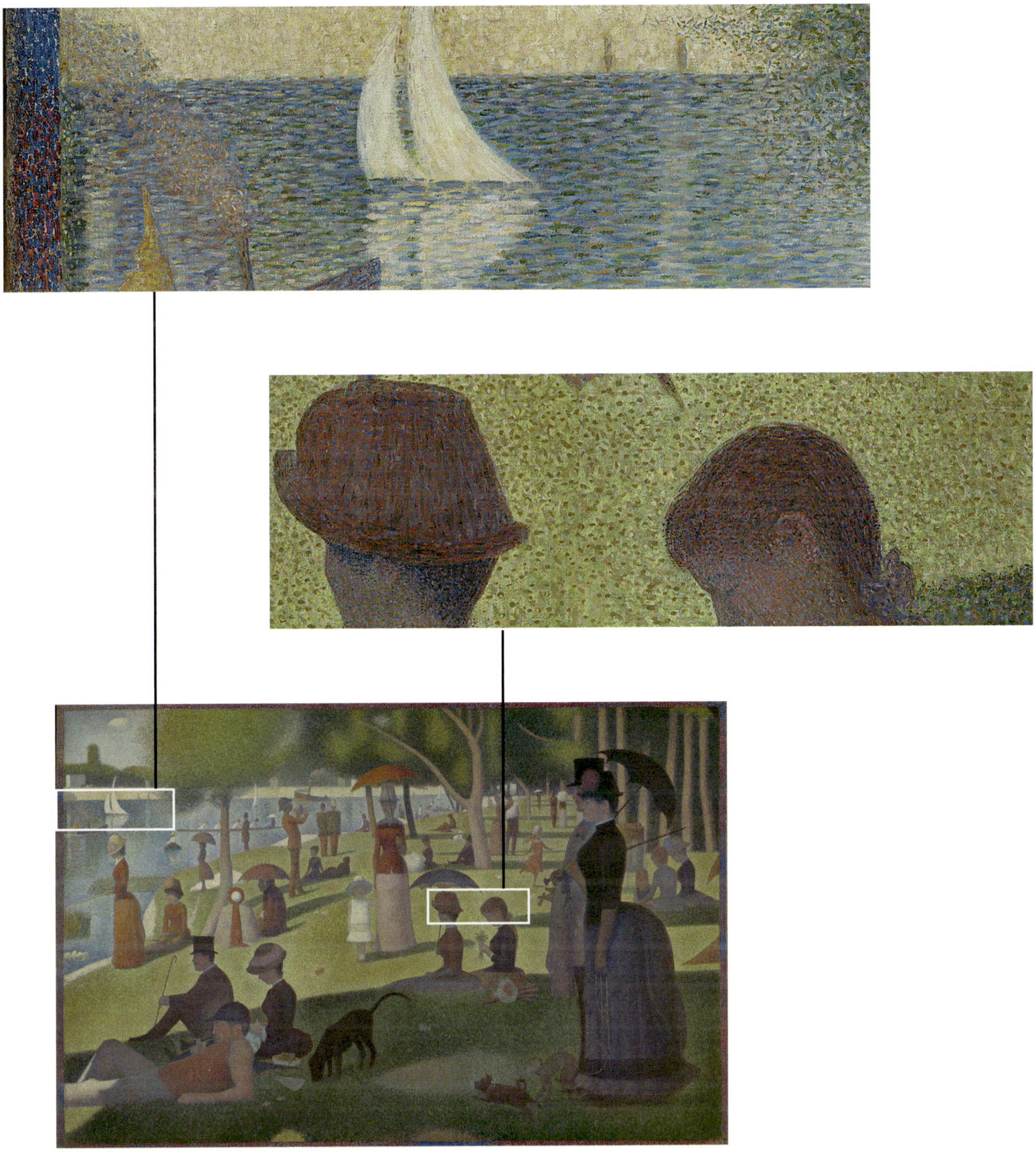

FIGURE 6.16. Georges Seurat (French, 1859–1891), *A Sunday on La Grande Jatte—1884*. 1884/86. Oil on canvas, 205.7 × 307.3 cm (81¾ × 121¼ in.). Art Institute of Chicago, 1926.224. Details of the two delineated areas of the painting are shown above and in figure 6.17. Notice that Seurat has used impressionist brushwork for the sailboat and a combination of impressionist brushwork and pointillist dabs and dashes for the water. The painted frame has pointillist brushwork.

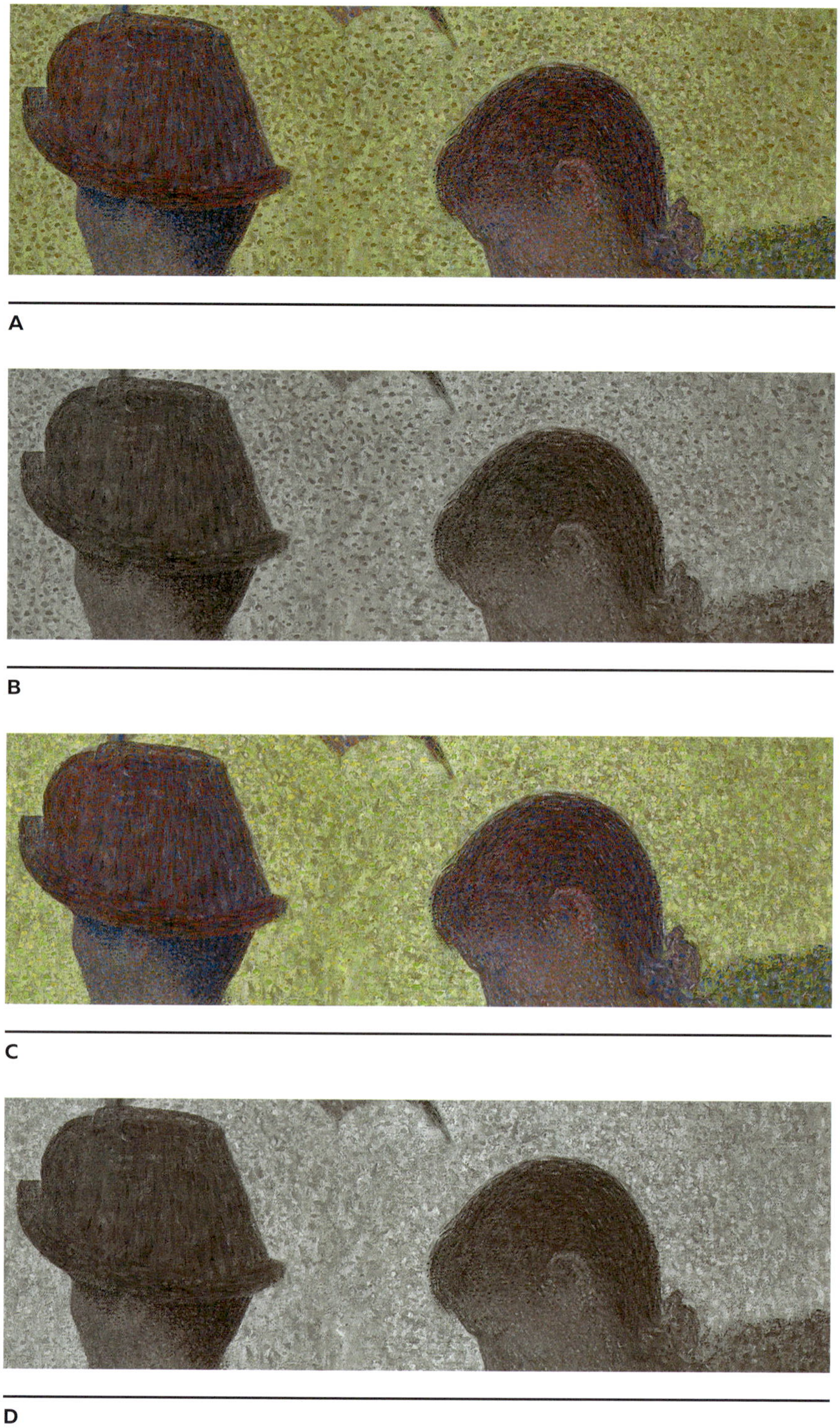

FIGURE 6.17. Detail of Seurat's *A Sunday on La Grande Jatte—1884*. The first two panels show the detail as (A) full color and (B) lightness, with the painting in its current condition. In the second two, the darkened zinc yellow has been computationally replaced with one that did not darken; the detail is again shown as (C) full color and (D) lightness.

light (Verhaeren 1891), came from dabs that had similar lightness and high chroma with large differences in hue. The constantly changing focus for these colors produced an illusion of vibration.[16] Because of the painting's immense size, differences in spatial vision in the center and periphery also contributed to the shimmering effect (Livingstone 2014).

Within several years, however, the zinc yellow in the painting had darkened, owing to light exposure, high humidity, and sulfur dioxide air pollution (Casadio, Fiedler, et al. 2008; Casadio, Xie, et al. 2011). With the vivid colors now turning to browns and olive greens, the painting lost much of its shimmering quality; the darkened dabs, now more visible against the lighter impressionist brushwork, gave the painting a more pointillist appearance. In figure 6.17, a demonstration of this process, the darkened zinc yellow was replaced computationally with a zinc yellow that did not darken, resulting in a dramatic change in the painting's appearance (Berns, Imai, et al. 2005; Berns, Byrns, et al. 2006). In this detail, the pointillist dabs containing nondarkening zinc yellow are less discernible, while the countershading around the right-hand figure has becomes more apparent.

Summary

Artist materials produce color by absorption, scattering, fluorescence, and interference. Pigments can be added that produce metallic and pearlescent effects, resulting in colors and spatial effects that change with illuminating and viewing geometry. This expanding choice of colorants will prove a challenge for conservation treatments such as imitative inpainting. We tend to think color is determined by differences in absorption as a function of wavelength. Scattering, however, is an equally important determinant of color, affecting vividness and depth. Differences in the opacity of a coloring material result from differences in scattering, which determines whether we will see the result of a buildup of colored layers or only the topmost layer. Picture varnishes change surface roughness. A smoother surface will increase the color gamut and sharpen the underlying paint layers. Colors can be mixed on the palette or on the painting using spatial patterns. In the latter case, colors are distance dependent—that is, the colors we see depend on whether the pattern can be resolved by our visual system at a particular distance from the painting.

Notes

1. The term "colorant" is used, rather than "pigment" or "dye," because a colorant can have the physical properties of either one, depending on the specific instance. For example, colorants that were historically considered pigments, such as earth oxides, can have negligible scattering if particle size is reduced, as in the case of transparent red oxide.
2. The specific paints used in this chapter are rutile titanium dioxide white (PW 6), carbon black (PBk 7), cobalt blue (PB 28), ultramarine blue (PB 29), phthalocyanine blue (PB 15:4 green shade), phthalocyanine green (PG 7), permanent green light (PY 3 / PG 7), pyrrole orange (PO 73), arylide yellow (PY 74), pyrrole red (PR 254), dioxazine purple (PV 23), and quinacridone magenta (PR 122) acrylic dispersion paints. Simplifications of their names will be used whenever possible.
3. Because arylide yellow's transition between shorter and longer wavelengths is not vertical, the intersecting wavelength also changes to a longer wavelength with positive shifts in the blue. However, this has a small effect compared with increasing maximum reflectance.
4. This is known as first-surface reflection and is a result of the refractive index discontinuity between air and the surface of the paint. For most paints, this is about 4% of the incident light.
5. Differences in refractive index (RI) between a paint layer and varnish layer will also affect the total amount of light reflected at a surface caused by interreflections at the paint-and-varnish and varnish-and-air interfaces. However, these interreflections are negligible compared with the effects caused by changes in surface roughness.
6. Low-molecular-weight resins smooth rough surfaces. Examples include dammar, mastic, Regalrez 1094, and MS2A. High-molecular-weight resins such as polyvinyl acetate have a negligible effect on a surface's roughness.
7. Differences in absorption at the short end of the visible spectrum can be used to identify different white colorants, including rutile and anatase titanium dioxide, lead, and zinc white.
8. These paints were simulated computationally by shifting the spectrum of arylide yellow toward either shorter or longer wavelengths.
9. Schott filter glass BG36 was used.
10. The experiment can also be performed by looking through the glass. In this case, all yellow-appearing wavelengths are removed from the environment.
11. The specular angle is defined as the angle of reflection equal and opposite to the angle of incidence.
12. The correct term is "near-aspecular," where aspecular uses the specular direction as the reference geometry.
13. Our perception of whiteness is quite interesting: decreasing yellowness has a more pronounced effect on increasing whiteness than does increasing lightness without a change in yellowness.

14. When fluorescent materials are measured, the units are total radiance factor rather than reflectance factor because emission and reflection are combined.
15. The *tratteggio* is darker than the adjacent gold. Since the lighting geometry is different in the imaging and conservation departments, a lightness mismatch occurs, causing the gold to lighten. This can also be seen when viewing the painting from different angles.
16. The change in focus, known as chromostereopsis, is a result of chromatic aberration in our visual system.

References

Barton, C. 2009. *The Day-Glo Brothers*. Watertown, MA: Charlesbridge.

Berns, R. S., S. Byrns, F. Casadio, I. Fiedler, C. Gallagher, F. H. Imai, A. Newman, and L. A. Taplin. 2006. "Rejuvenating the Color Palette of Georges Seurat's 'A Sunday on La Grande Jatte—1884': A Simulation." *Color Research and Application* 31: 278–93.

Berns, R. S., and E. R. de la Rie. 2003a. "Exploring the Optical Properties of Picture Varnishes Using Imaging Techniques." *Studies in Conservation* 48: 73–82.

——. 2003b. "The Effect of the Refractive Index of a Varnish on the Appearance of Oil Paintings." *Studies in Conservation* 48: 251–62.

Berns, R. S., F. H. Imai, L. A. Taplin, S. Byrns, F. Casadio, I. Fiedler, C. Gallagher, and A. Newman. 2005. "Rejuvenating the Appearance of Seurat's 'A Sunday on La Grande Jatte—1884' Using Colour and Imaging Science Techniques: A Simulation." In *14th Triennial Meeting, The Hague, 12–16 September 2005: Preprints (ICOM Committee for Conservation),* edited by I. Verger, 452–58. London: Earthscan.

Berns, R. S., and M. Mohammadi. 2007. "Evaluating Single- and Two-Constant Kubelka-Munk Turbid Media Theory for Instrumental-Based Inpainting." *Studies in Conservation* 52: 299–314.

Casadio, F., I. Fiedler, K. A. Gray, and R. Warta. 2008. "Deterioration of Zinc Potassium Chromate Pigments: Elucidating the Effects of Paint Composition and Environmental Conditions on Chromatic Alteration." In *ICOM-CC 15th Triennial Conference Preprints, New Delhi, 22–26 September 2008,* edited by J. Bridgland, 572–80. Paris: International Council of Museums.

Casadio, F., S. Xie, S. C. Rukes, B. Myers, K. A. Gray, R. Warta, and I. Fiedler. 2011. "Electron Energy Loss Spectroscopy Elucidates the Elusive Darkening of Zinc Potassium Chromate in Georges Seurat's 'A Sunday on La Grande Jatte—1884.'" *Analytical and Bioanalytical Chemistry* 399: 2909–20.

Cotte, P., and D. Dupraz. 2006. "Multispectral Photography of the Famous Mona Lisa Painting." CGIV 2006 3rd European Conference on Colour in Graphics, Imaging, and Vision, Leeds, UK.

de la Rie, E. R. 1987. "The Influence of Varnishes on the Appearance of Paintings." *Studies in Conservation* 32: 1–13.

——. 1988. "Photochemical and Thermal Degradation of Films of Dammar Resin." *Studies in Conservation* 33: 53–70.

——. 1989. "Old Master Paintings: A Study of the Varnish Problem." *Analytical Chemistry* 61: 1228A–33A, 1237A–40A.

Dekker, N., E. J. J. Kirchner, R. Supèr, G.-J. van den Kieboom, and R. Gottenbos. 2011. "Total Appearance Differences for Metallic and Pearlescent Materials: Contributions from Color and Texture." *Color Research and Application* 36: 4–14.

Elias, M., and P. Cotte. 2008. "Multispectral Camera and Radiative Transfer Equation Used to Depict Leonardo's *sfumato* in *Mona Lisa*." *Applied Optics* 47: 2146–54.

Ellis, M. H., C. W. McGlinchey, and E. Chao. 2002. "Daylight Fluorescent Colors as Artistic Media." In *The Broad Spectrum: Studies in the Materials, Techniques, and Conservation of Color on Paper,* edited by H. K. Stratis and B. Salvesen, 160–66. London: Archetype.

Fiedler, I. 1984. "Materials Used in Seurat's *La Grande Jatte,* Including Color Changes and Notes on the Evolution of the Artist's Palette." In *AIC Preprints, American Institute for Conservation 12th Annual Meeting, Los Angeles,* 43–51. Washington, DC: AIC.

Herbert, R. L., and N. Harris. 2004. *Seurat and the Making of "La Grande Jatte."* 1st ed. Chicago: Art Institute of Chicago and University of California Press.

Johnston-Feller, R. 2001. *Color Science in the Examination of Museum Objects: Nondestructive Procedures,* Tools for Conservation. Los Angeles: Getty Conservation Institute.

Kirchner, E., G.-J. van den Kieboom, L. Njo, R. Supèr, and R. Gottenbos. 2007. "Observation of Visual Texture of Metallic and Pearlescent Materials." *Color Research and Application* 32: 256–66.

Kuehni, R. G. 2012. "Unique Hues and Their Stimuli—State of the Art." *Color Research and Application* 39: 279–87.

Livingstone, M. 2014. *Vision and Art: The Biology of Seeing*. Updated and expanded edition. New York: Harry N. Abrams.

Mohen, J.-P., M. Menu, and B. Mottin. 2006. *Mona Lisa: Inside the Painting*. New York: Harry N. Abrams.

Morris, H. R., and P. M. Whitmore. 2007. "'Virtual Fading' of Art Objects: Simulating the Future Fading of Artifacts by Visualizing Micro-Fading Test Results." *Journal of the American Institute for Conservation* 46: 215–28.

Mustalish, R. A. 2000. "Optical Brighteners: History and Technology." In *Tradition and Innovation: Advances in Conservation; Contributions to the Melbourne Congress, 10–14 October 2000,* edited by A. Roy and P. Smith, 133–36. London: International Institute for Conservation of Historic and Artistic Works.

Nadolny, J. 2012. "History of Visual Compensation for Paintings." In *The Conservation of Easel Paintings*, edited by J. H. Stoner and R. Rushfield, 573–85. New York: Routledge.
Pfaff, G. 2008. *Special Effect Pigments*. Hanover: Vincentz Network.
Rood, O. N. 1879. *Modern Chromatics, with Applications to Art and Industry*. International Scientific Series 27. New York: D. Appleton.
Verhaeren, E. L. 1891. "Georges Seurat." *La Société, Nouvelle* 7 (Apr.): 1.

COLOR REPRODUCTION

7

Producing images and reproductions of artwork is indispensable for the creators, conservators, and connoisseurs of works of art. Artists often keep records of their work, and they may want to make signed prints or posters from an original. Conservators, for their part, need to document condition and any treatments they undertake. Images created using a variety of technologies have become an important tool in the technical examination of art. Connoisseurs often use reproductions as surrogates for the actual object.

The availability of art images in digital format seems boundless. Often, unfortunately, the range of colors of these images is also boundless, as an image search quickly reveals. Different imaging systems produce different colors because of inherent differences in cameras and lighting. Many images are not color managed, and as a result our computers misinterpret their RGB data. Printing can also distort color. In some cases the color gamut achievable for a specific set of inks and paper type is too small to reproduce a painting's vivid colors. A final possibility is that an author or publisher may have wanted the printed image to have a certain aesthetic; the intention may not have been to reproduce colors accurately.

Many printing technologies use four inks—cyan, magenta, yellow, and black (CMYK). Here these are visible in the form of magnified dots in a detail of Renoir's *Albert Cahen d'Anvers*. Modulating the size of the dot within a fixed pattern produces full color (detail, fig. 7.2f).

Goals of Color Reproduction

Three color-reproduction goals are relevant to artwork reproduction: preferred, spectral, and, colorimetric. Chemical photography is an excellent example of preferred color reproduction. As color photography evolved, it became clear that consumers did not want a color-accurate recording of an event; rather, they wanted a picture that was both beautiful and matched their memory of the event, achieved by increasing the chroma of familiar objects, such as green grass and blue sky, and increasing overall lightness contrast (Newhall, Burnham, and Clark 1957; Hunt, Pitt, and Winter 1974). Preferred color reproduction was well understood by the middle of the twentieth century (Evans 1959; Hunt 1970), and as digital photography evolved, image processing was used to mimic chemical photography. This still occurs today, from mobile devices to medium-format studio cameras.

The goal of spectral color reproduction is to match the spectral reflectance properties of the artwork. This requires a multi- or hyperspectral camera and a printing system whose ink and paper can match the artwork's spectral properties. Spectral color reproduction has two advantages. First, the reproduction matches the original under any lighting condition and when viewed by any observer. The print will match the painting both in the imaging studio and in the gallery. Second, the painting and the print will have the same color inconstancy. By taking the print outdoors, we can see how the painting appeared to a *plein air* painter. The spectral match is also a colorimetric match, with CIELAB coordinates for both the painting and the print that are identical for all observers and all lighting conditions.

The vast majority of digital cameras are RGB; that is, they have three channels to record a painting's colors. With just three channels, it is possible to match XYZ tristimulus values or CIELAB coordinates for only a single observer and a single type of lighting. This is known as colorimetric color reproduction.

To create figure 7.1, two different systems—preferred and colorimetric—were used to image Paul Cézanne's *Portrait of Anthony Valabrègue*. The preferred system used a scanner to digitize a photographic transparency, resulting in high lightness and color contrast, appearing as if overcleaned. The colorimetric image is a more accurate record of the painting's colors, which are typical of a mid-nineteenth-century painting with a natural varnish about thirty years old.

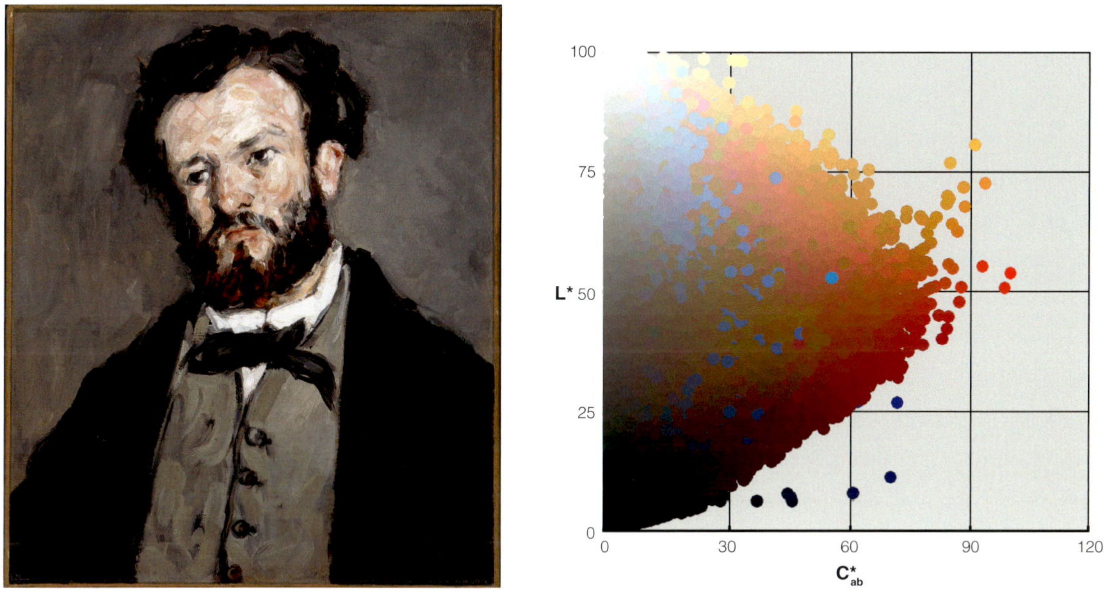

A

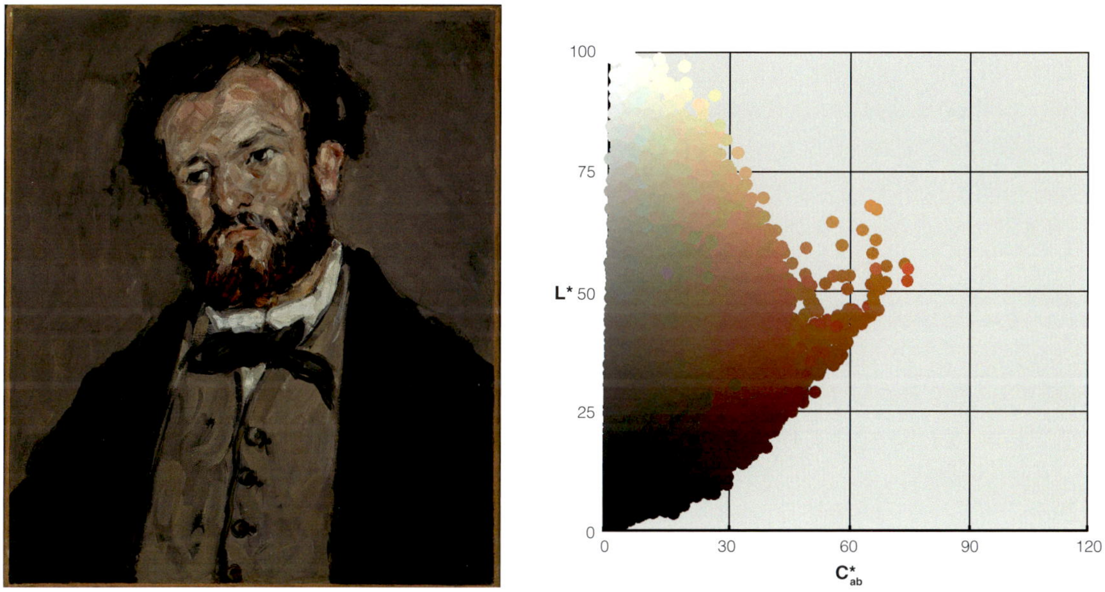

B

FIGURE 7.1. Paul Cézanne (French, 1839–1906), *Portrait of Anthony Valabrègue*, ca. 1870. Oil on canvas, 60 × 50.2 cm (23⅝ × 19¾ in.). Los Angeles, J. Paul Getty Museum, 85.PA.45. Principles of either (A) preferred or (B) colorimetric color reproduction are used to make each image, with its colors plotted in CIELAB to the right. The green and blue colors correspond to image noise, observable only at high magnification.

The Image

The properties of a digital image include dimensions, resolution, bit depth, file type, color mode, and color encoding (color management). We can think of a digital image as a piece of graph paper. Each square is called a pixel (a shortened form of "picture element"), and the number of squares on the sheet of paper defines the resolution. Digital cameras are marketed according to their total number of pixels. A 24-megapixel (mpixel or MP or M) camera has 24 million pixels, usually arranged as 6,000 pixels in one direction and 4,000 pixels in the other direction: 6,000 x 4,000 = 24,000,000. As consumers, we have been trained to equate the number of pixels with image quality. But there is a more critical dimension: the size of the sensor. From the number of pixels and the sensor size, the size of each pixel can be determined. Larger pixels have better noise properties. This is why studio cameras have larger sensors than consumer cameras. A small-megapixel camera can easily outperform one with many more pixels if it has superior optics and a sensor area that is the same size or larger. As these factors are more important than the total number of pixels, the images will be sharper and lower in noise. Note also that using Adobe Photoshop to increase resolution does not improve image quality, just file size.[1]

Our numerical system is the decimal system, based on powers of 10 organized in ascending columns from right to left. Each column has ten unique values: 0, 1, 2, . . ., 9. Images use the binary system, based on powers of 2. Each position has two unique values: 0 and 1. Each column is called a bit. Most commonly, images use 8-bit encoding with values ranging from 00000000 to 11111111. There are 256 unique values, 0 to 255. The number of bits, called bit depth, determines numerical precision. Previously, most imaging hardware and software was limited to 8 bits; this precision was just adequate if best practices were followed, particularly during image capture and visual editing. Recently, digital cameras have advanced to the point where their precision exceeds 12 bits, and improvements continue to be made. An 8-bit-per-channel image (24 total bits for RGB) will negate such improvement, and 16- and 32-bit precisions are now used. If an image is a legacy 8-bit-per-channel image or if a digital camera stores only 8-bit-per-channel images, changing the bit depth to 16 or 32 will not improve image quality, just increase its file size.[2]

Two common file types are TIFF (image.tif or image.tiff) and JPEG (image.jpg). TIFF files contain complete data, whereas JPEG files have reduced data that does not degrade image quality for many viewing applications. This type of data reduction is known as visually lossless file compression. Most amateur photographers store images in JPEG format, whereas professionals store images in TIFF format or a raw image format that is specific to the camera manufacturer. JPEG image compression should not be used for scientific imaging or for archiving artwork.

The color mode defines the number of channels. Modes include grayscale (one channel), RGB (three channels: red, green, and blue), CMYK (four channels: cyan, magenta, yellow, and black),[3] and Lab (three channels: L*, a*, and b*). A color mode is underspecified until the color coordinates are related to a set of reference conditions. This process, called color encoding, is the domain of color management. The exception is a Lab image. When an image is assigned as a Lab image, it is encoded as CIELAB L*, a*, and b* for CIE illuminant D50 and the 1931 standard observer.

Image Quality

Producing a color image can be as simple as pointing a mobile phone at the artwork or as complicated as the practices of a museum's imaging department (Berns, Frey, et al. 2005; Frey and Farnard 2011). How do we determine quality requirements? How is quality even defined? A good place to start is to look at images with different kinds of image-quality limitations.

Quality is separated into spatial and color quality, visualized in figures 7.2 and 7.3. Spatial quality, explored in figure 7.2 with a detail of an eye from Renoir's *Albert Cahen d'Anvers,* can be evaluated visually by looking at images at full resolution on screen ("100% magnification" in Photoshop). A silkscreened or paint-by-numbers appearance may come from insufficient bit depth (fig. 7.2b). The problem can also arise when an underexposed image is rescaled ("levels adjustment" in Photoshop). Limitations in the optical properties of lenses, the need to have a shallow depth of field sometimes, or an inability to focus can result in image blur (fig. 7.2c). This can be corrected to a large extent by sharpening, which involves amplifying the high spatial frequencies, but there is the danger of oversharpening, with unnatural edge contrast as a result (fig. 7.2d). Another problem with quality stems from the uncertainty that is part of all processes. In imaging, uncertainty takes the form of a random variability called image noise, which gives a speckled appearance (fig. 7.2e). Depending on the cause of the noise, the speckle may appear chromatic or achromatic

FIGURE 7.2. (A) Detail of Renoir's *Albert Cahen d'Anvers*, with examples of (B) insufficient number of levels, (C) blurring, (D) oversharpening, (E) excessive random noise, and (F) systematic pattern noise.

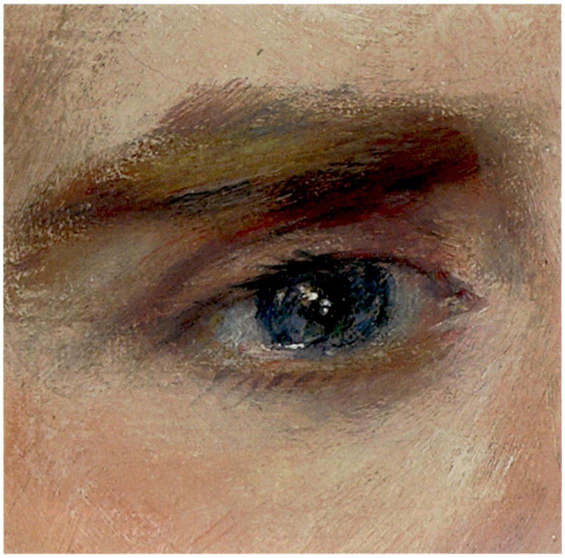

A

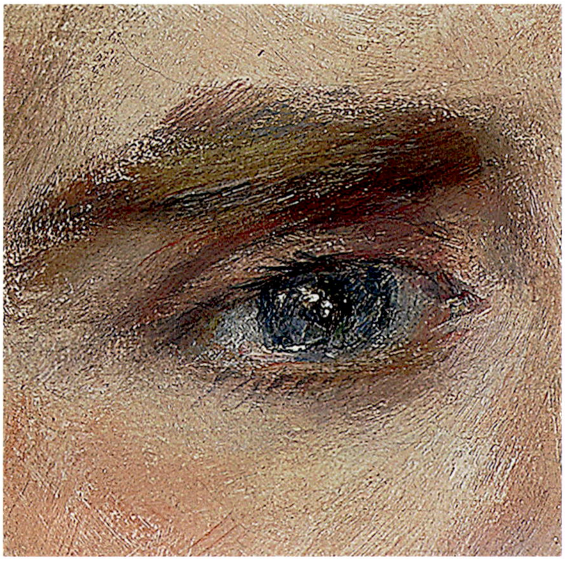

D

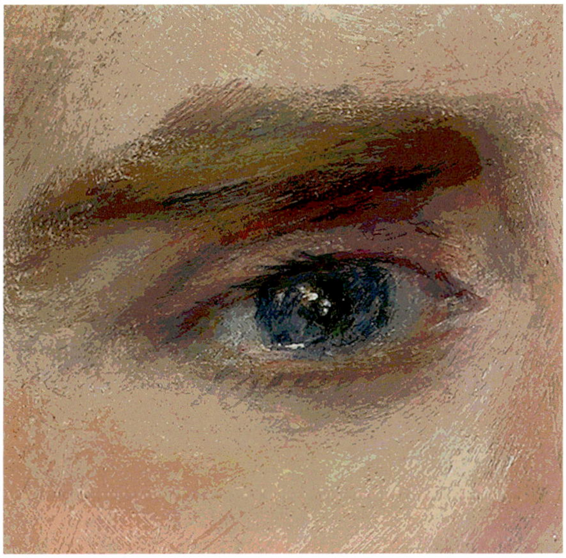

B

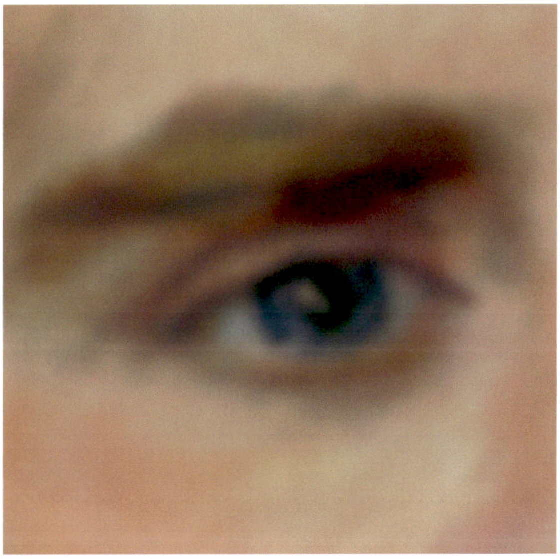

C

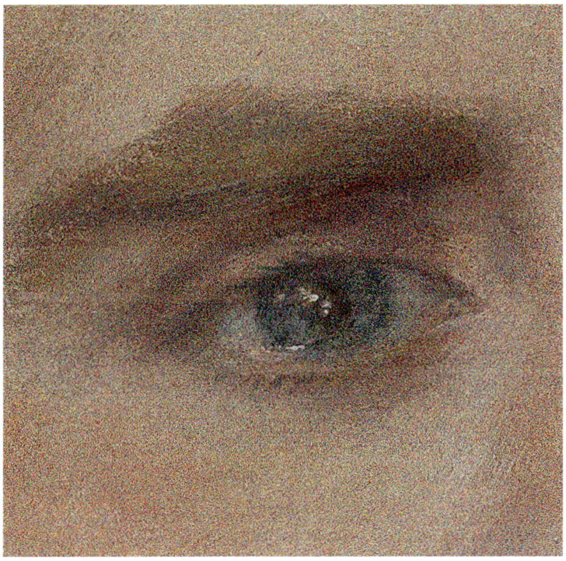

E

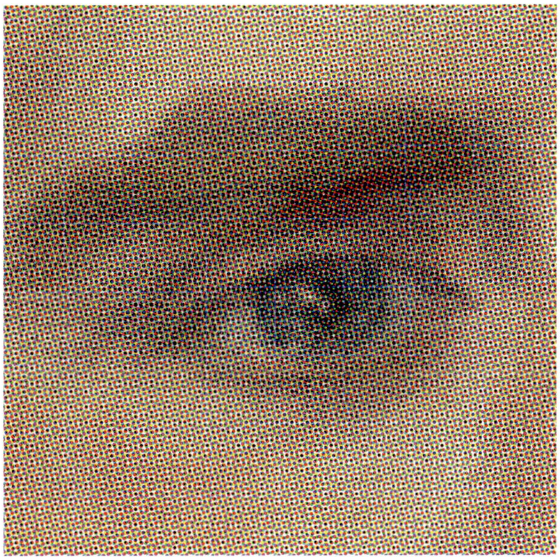

F

FIGURE 7.3. (A) Detail of Renoir's *Albert Cahen d'Anvers,* with examples of (B) color balance errors, (C) lightness (tone reproduction) errors, and (D) chroma errors.

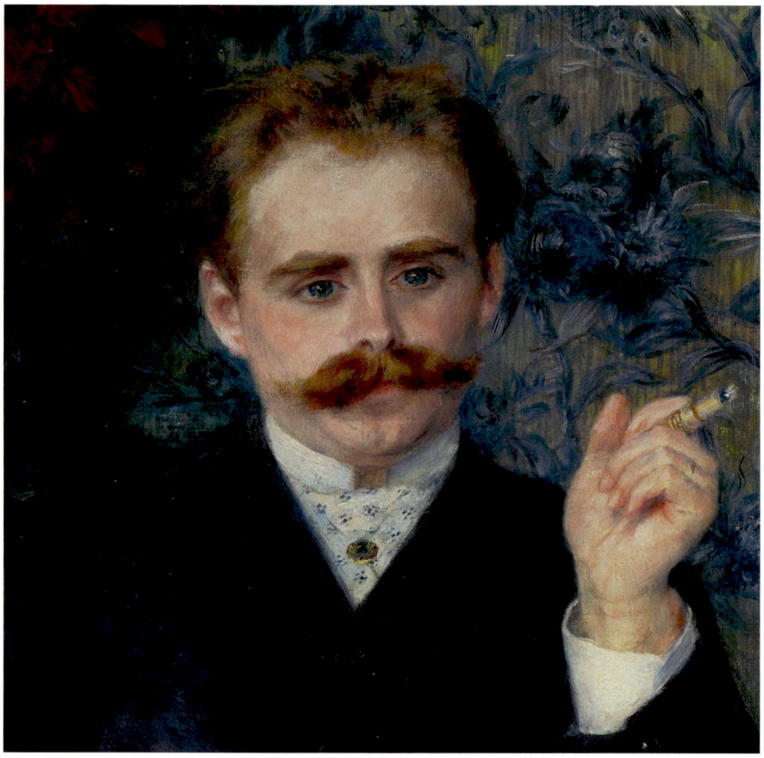

A

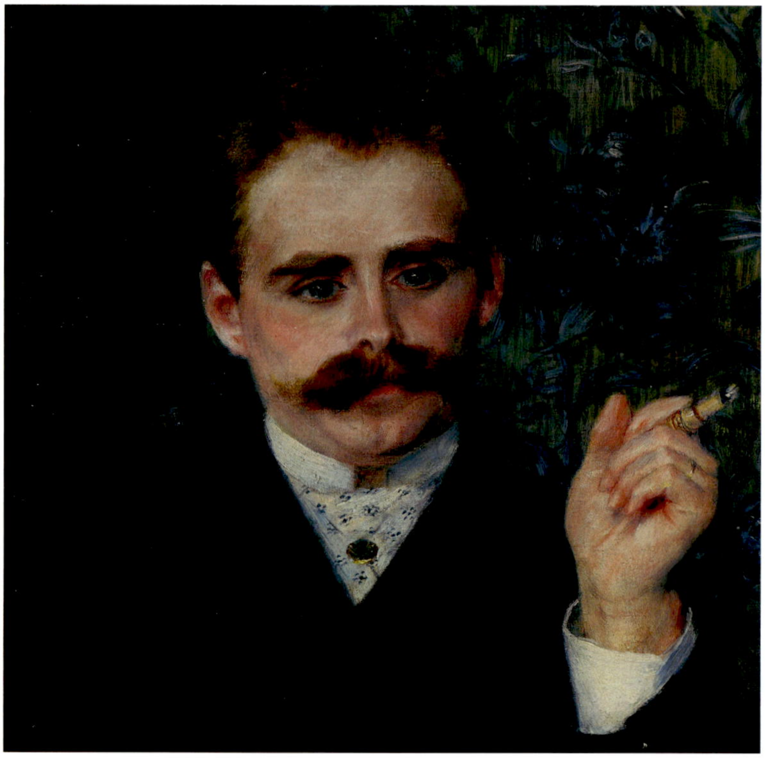

C

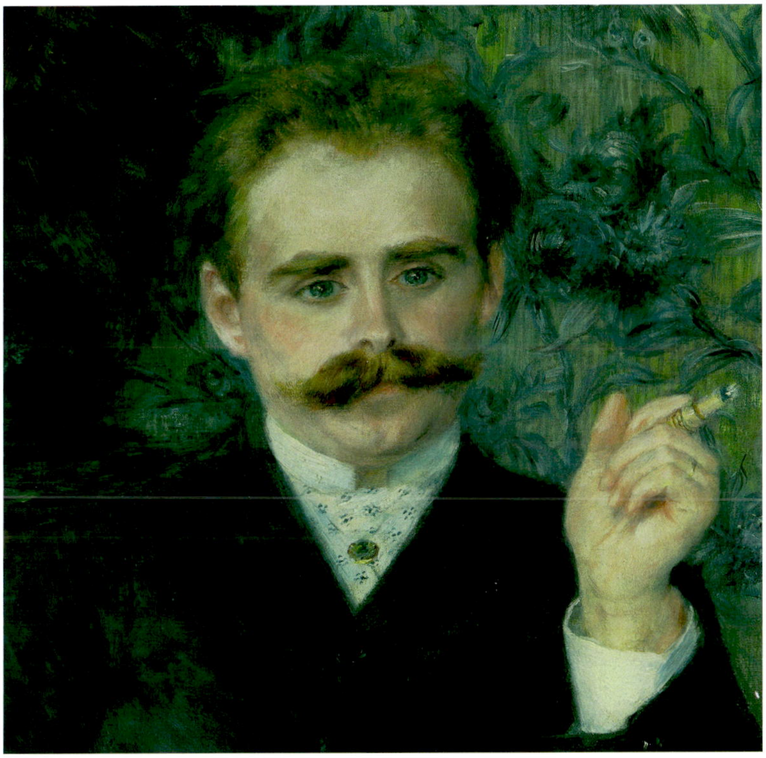

B

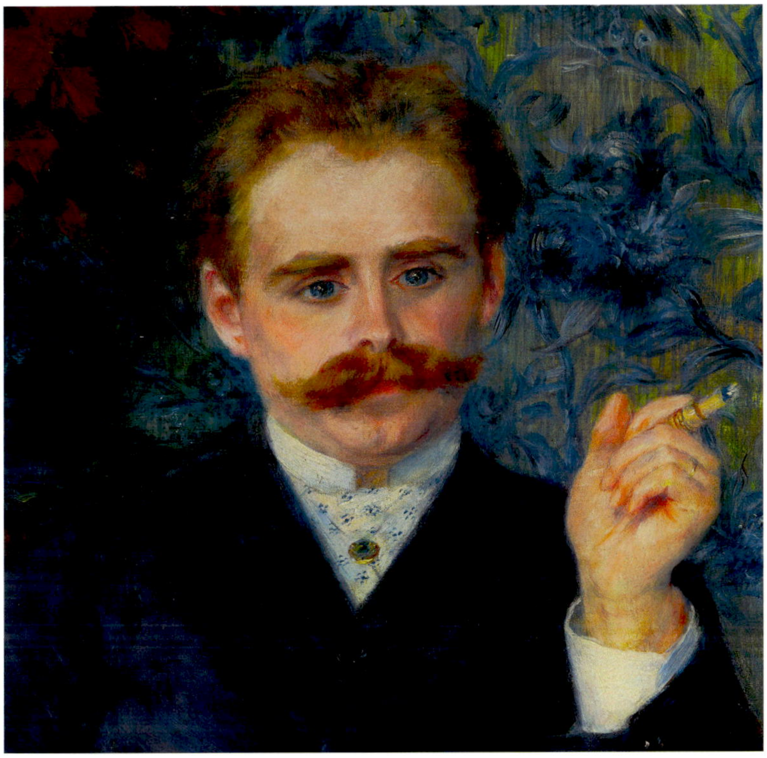

D

and is often observable in dark colors. Finally, there may be a spatial pattern, as in the case of low-quality printing (fig. 7.2f). The visibility of spatial image-quality limitations depends on the particular spatial image content of the painting, the viewing distance, and the image magnification.

Color quality, explored in figure 7.3 with a larger detail from Renoir's painting, is first evaluated by looking at color balance. Consumer imaging attempts to mimic our chromatic adaptation, including incomplete adaptation: think of the warm color balance of the candlelit dinner. When artwork is imaged, adaptation is assumed to be complete, and any lighting-based color cast should be removed (fig. 7.3b). Next, lightness accuracy is evaluated; this is often called tone reproduction. The most common error is a mismatch in the middle tones (fig. 7.3c): the values for white and black are correct, but everything in between is not. Chroma errors occur when the image is boosted using Photoshop's levels, contrast, saturation, and vibrance adjustment tools (fig. 7.3d). When increases in chroma are too extreme, the image appears unnatural (Yendrikhovskij, Blommaert, and de Ridder 1999), though this can be difficult to judge when viewing images of artwork. Hue errors can also occur. In this case, only specific colors have incorrect hue, not the entire image. The main causes of color errors of all kinds are poor color management, inappropriate visual editing, and fundamental limitations in the color engineering of imaging devices.

Color Management

Before digital imaging was ubiquitous, color reproduction was the domain of professionals using specialized equipment. We sent our rolls of film out for processing and received either transparencies or prints. Photographs and text were sent to printers, and books were sent back. In the 1980s, developments in desktop computers, color displays, color scanners, digital cameras, color printers, and WYSIWYG ("what you see is what you get") software launched digital color imaging and a new approach to color reproduction—desktop publishing. Chemical photography was no longer required to produce colored images. Images could be viewed on a display as well as in the form of prints. Changes in the printing industry were even more dramatic. Suddenly, many of the functions performed by highly trained specialists, using expensive technology such as analog scanning and color separation, could be performed digitally using equipment that was much lower in cost.

After the initial excitement of being able to create, view, and print colored images had worn off, many users became dissatisfied with inconsistent color and, often, low color quality. The solution was to manage color through reference color-imaging systems: one for desktop publishing and, several years later, one for consumer imaging. The consumer reference system was a color cathode ray tube (CRT) driven by a computer running Microsoft's operating system, that is, a PC. Based on measurements of PCs in office settings and existing television and computer-controlled display standards, a reference display was defined – sRGB, the "s" being an abbreviation for "standard" (IEC 1999). Cameras and scanners would define images using sRGB. These images would appear reasonable when displayed directly on a PC. Printers became sRGB devices where internal hardware and software transformed RGB data into CMYK signals that controlled the printer. sRGB cameras could be plugged directly into sRGB printers. The sRGB workflow was very effective in improving color consistency, and when image processing was optimized for preferred color reproduction, the resulting images met most consumer expectations.

Desktop publishing required a reference color-imaging system with a much larger color gamut than a CRT display. Imaging scientists and engineers also recognized that defining images using a device's native internal data would improve spatial image quality for cameras, scanners, and displays by reducing effects caused by insufficient bit depth and would be better suited for printing systems like offset printing that defined colors only by CMYK. Rather than working through a standards organization such as the CIE, in 1993 the leading imaging companies of the day formed the International Color Consortium, ICC (Green 2010).[4] This consortium standardized a reference system and protocol for translating native data into and out of the reference system.

The basic workflow for the ICC's color-management system is shown in figure 7.4. Each imaging device has its own native digital color signals. The profile converts native device data to data controlling the reference color-imaging system. The profile is assigned to an image as a component of non-image data called metadata. Device data can be thought of as forming a color space: RGB is a three-dimensional space, and CMYK is four-dimensional. Using this terminology, the reference-imaging-system data form what is called the profile connection space (PCS), so called because the reference imaging system connects different imaging

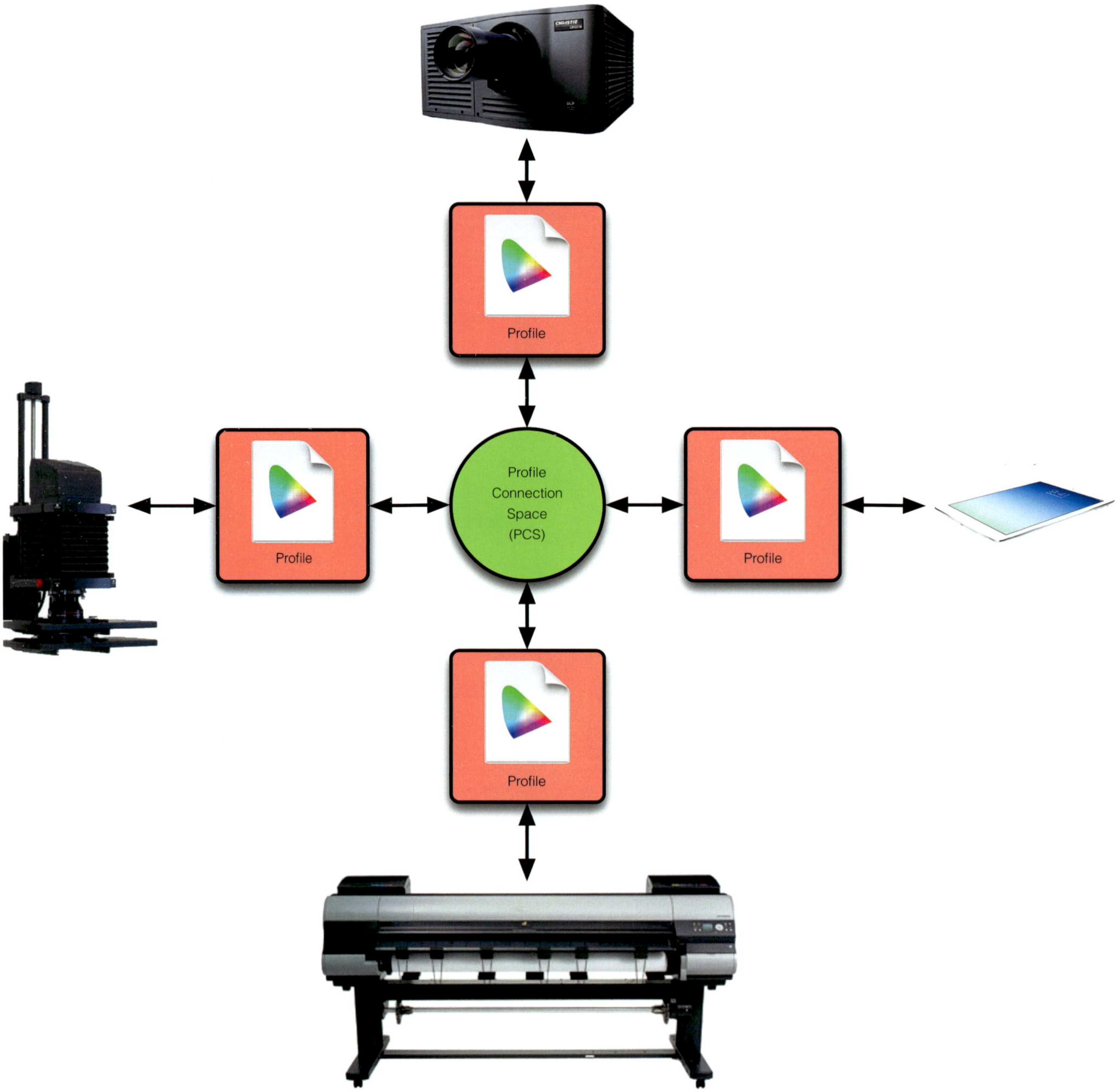

FIGURE 7.4. ICC color-management workflow diagram. Profiles are specific to each device. If each device produces the same PCS coordinates, the colors match.

devices. The PCS defines the digital signals that control an ideal printer, with paper that reflects 100% of the incident light and colorants that produce an extremely large color gamut.[5] This ideal print is illuminated by CIE illuminant D50 and viewed by the CIE 1931 standard observer. This reference imaging system is further abstracted to a colorimetric specification, either XYZ or Lab, leading back to the tenets of colorimetry and colorimetric color reproduction: two colors (or images) that have the same colorimetry match each other for a specific illuminant and specific observer.

Software known as the color engine or color-management module (CMM) is used to convert between devices and the PCS. In practice, profiles are concatenated, and there is a direct conversion between imaging devices: an RGB image is converted to CMYK. The most common software to implement ICC-based color management is Adobe Photoshop. In fact, Adobe was one of the founding members of the ICC. Today, Adobe products control their colors using ICC conventions.

Accurate profiles are the key to effective color management. If a profile is created for bright-white paper and printed on brownish yellow newsprint, the resulting colors will not look correct. A profile created for newsprint will compensate for the paper color. Every imaging device has one or more default profiles: cameras, scanners, and displays have a single profile while printers have multiple profiles, one for each kind of paper. It is also possible to create and optimize profiles using independent software and hardware. The most effective profiling systems use measurement instrumentation — for example, a spectrophotometer when profiling a printer and a spectroradiometer when profiling a display. This way, we are assured that the profile corresponds to the actual conditions.

Most consumer cameras encode using sRGB even though CRT technology is nearly obsolete. If an image does not have a profile, it is safest to assume sRGB encoding. However, if there is a profile and the viewing software does not recognize it — as occurs in some web browsers — and assumes sRGB encoding, severe color errors like those shown in figure 7.5 can result.

Professional camera systems use a variety of encoding spaces. The most common are sRGB, Adobe RGB (1998), eciRGB v2, and ProPhoto RGB. The particular choice depends on bit depth, existing standards and recommendations, the use of visual editing software, and the scene colors. For museum imaging, ProPhoto RGB with 16-bit depth is a good choice

FIGURE 7.5.
(A) ProPhoto RGB image of Renoir's *Albert Cahen d'Anvers,* (B) rendered under the assumption that the image encoding is sRGB.

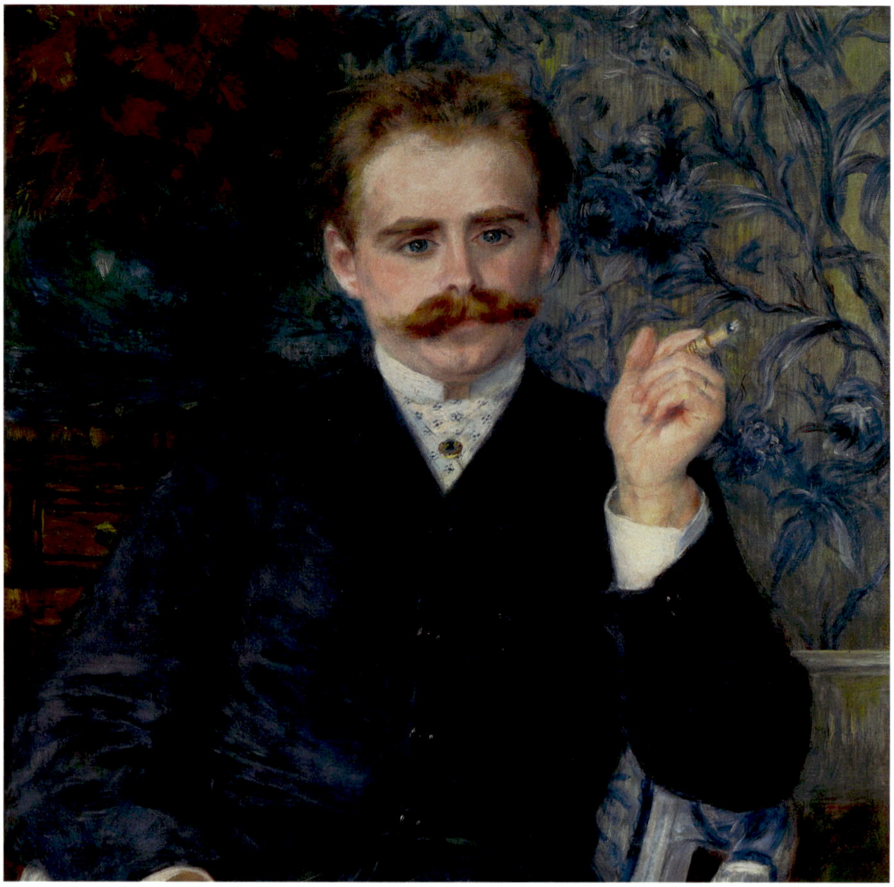

A

because artist paints are within its color-rendering gamut;[6] with their smaller rendering gamuts, all the other spaces produce encoding errors (Berns 2014a; Berns and Derhak 2014). The problem is demonstrated in figure 7.6 for mixtures of twenty-one artist acrylic dispersion paints with a glossy picture varnish.[7] Many of the colors cannot be encoded accurately for sRGB and Adobe RGB (1998), encoding spaces often used in museums.

Printers also contain profiles. Instead of residing within the metadata, the profiles are a component of the software used to set up and control the printer. It is up to the user to choose the correct profile for each paper type.

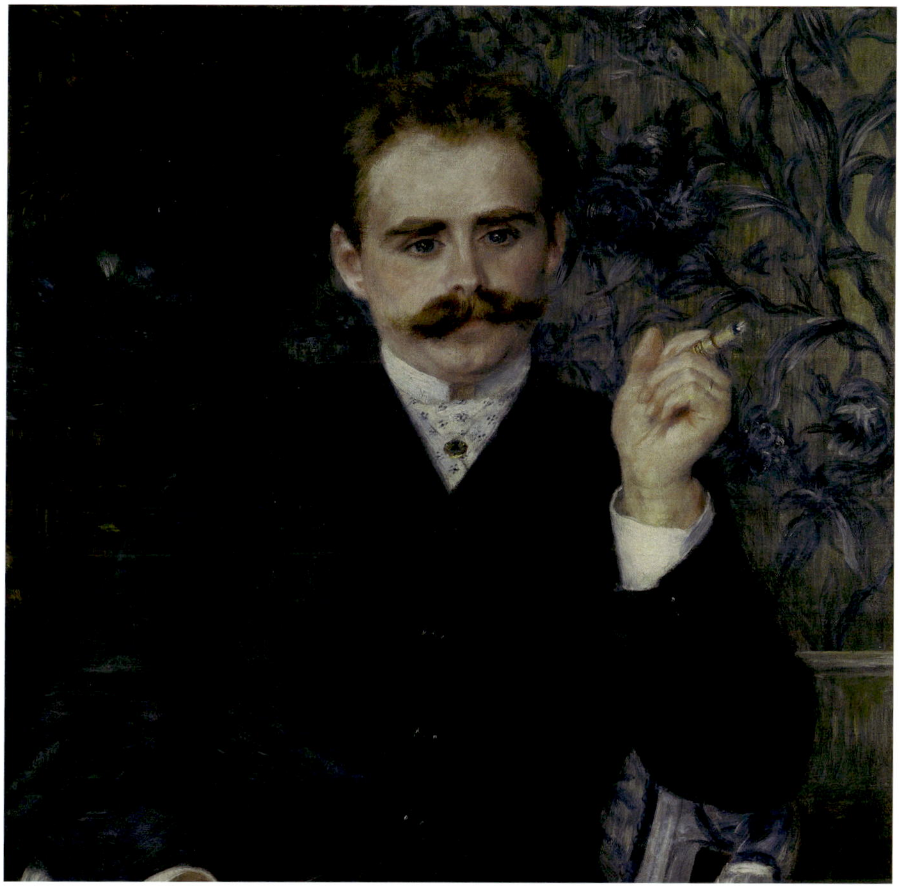

B

Once all the necessary profiles are in place, the user converts to a profile—the process of implementing color management, performed by the color engine. Suppose the image is very chromatic and many of the colors cannot be printed accurately. Rather than the paper remaining blank in those places, these colors are changed to fit within the printing system's color gamut. The opposite can also occur: an image with lackluster colors might be printed to take advantage of a system's color capabilities. There are many approaches to changes of this kind, which are known as color-gamut mapping (Morovic 2008). The ICC has defined four general categories, called color-rendering intents: perceptual, saturation, media-relative colorimetric, and ICC-absolute colorimetric.[8]

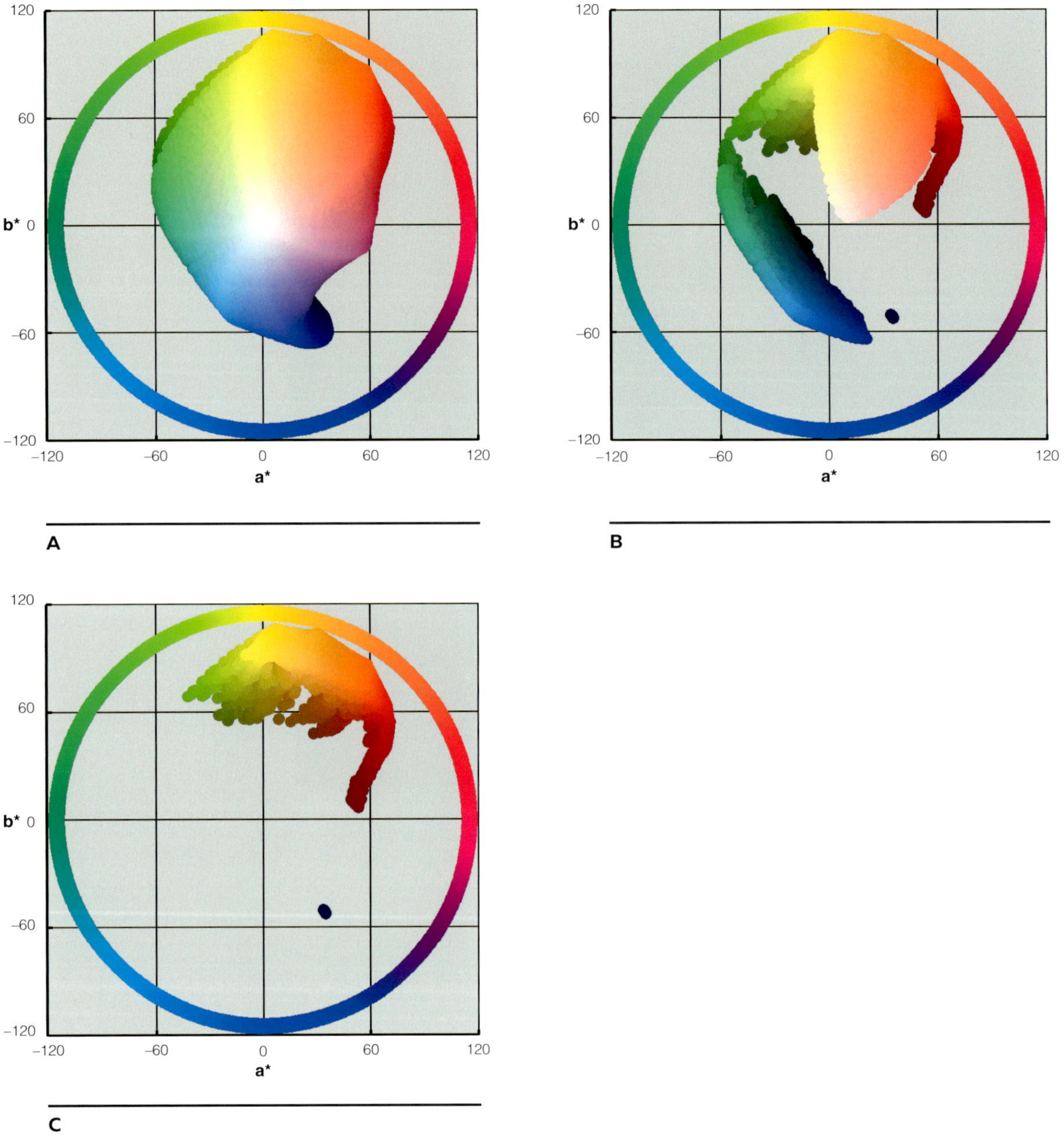

FIGURE 7.6.
(A) CIELAB coordinates of mixtures of twenty-one artist acrylic dispersion paints with a glossy picture varnish. Colors that cannot be encoded accurately—that is, out-of-gamut colors—are shown for (B) sRGB and (C) Adobe RGB (1998) encoding spaces.

Perceptual mapping mimics chemical photographic systems based on principles of preferred color reproduction. Such mapping also has the option of "black-point compensation," in which the black of the input is remapped to the black of the output. (Saturation and media-relative colorimetric also have the black-point compensation option.) Perceptual rendering can produce large errors in color and is not recommended for artwork reproduction.

Saturation mapping distorts colors to fully utilize a device's color gamut and should also not be used for artwork reproduction.

The media-relative colorimetric rendering, including black-point compensation, produces a color match on the assumption that each device's white defines the observer's state of chromatic adaptation. Because of the rescaling of the white points, the amount of clipping caused by differences in color gamut is greatly reduced. The same is true when rescaling the black points. Thus, media-relative colorimetric rendering intent with black-point compensation can be effective for artwork reproduction.

The ICC-absolute colorimetric mapping also produces a match for colors within the output device's color gamut and will clip any colors outside of the color gamut. Because it is not remapping the image white, this rendering intent can be effective for artwork with a slightly colored substrate, such as many early photographic prints.

Choosing between media-relative and ICC-absolute rendering intents when printing depends on the application. When a duplicate of the same size is being made, ICC-absolute rendering should be used. When the image is in a book where unprinted paper is visible around the image, this rendering can have an excessive color cast, and media-relative rendering may be a better choice. The two are compared in figure 7.7 for a painting by Édouard Manet.

Print profiles include all four rendering intents. It may be prudent to evaluate each one, with and without black-point compensation, by printing a digital target—as shown in figure 7.8—with known CIELAB values, measuring each sample with a spectrophotometer and calculating CIELAB, and then determining the best profile for a given usage based on the colorimetric data.

FIGURE 7.7.
Édouard Manet (French, 1832–1883), *Le Printemps (Jeanne Demarsy)*, 1881. Oil on canvas, 74 × 51.4 cm (29⅛ × 20¼ in.). Los Angeles, J. Paul Getty Museum, 2014.62. The Adobe RGB (1998) encoded image is rendered according to (A) media relative colorimetric intent, which compensates for the color of both the viewing illuminant and the paper, and (B) the ICC-absolute colorimetric intent, which preserves the encoded white point, resulting, in this instance, in a bluish color balance.

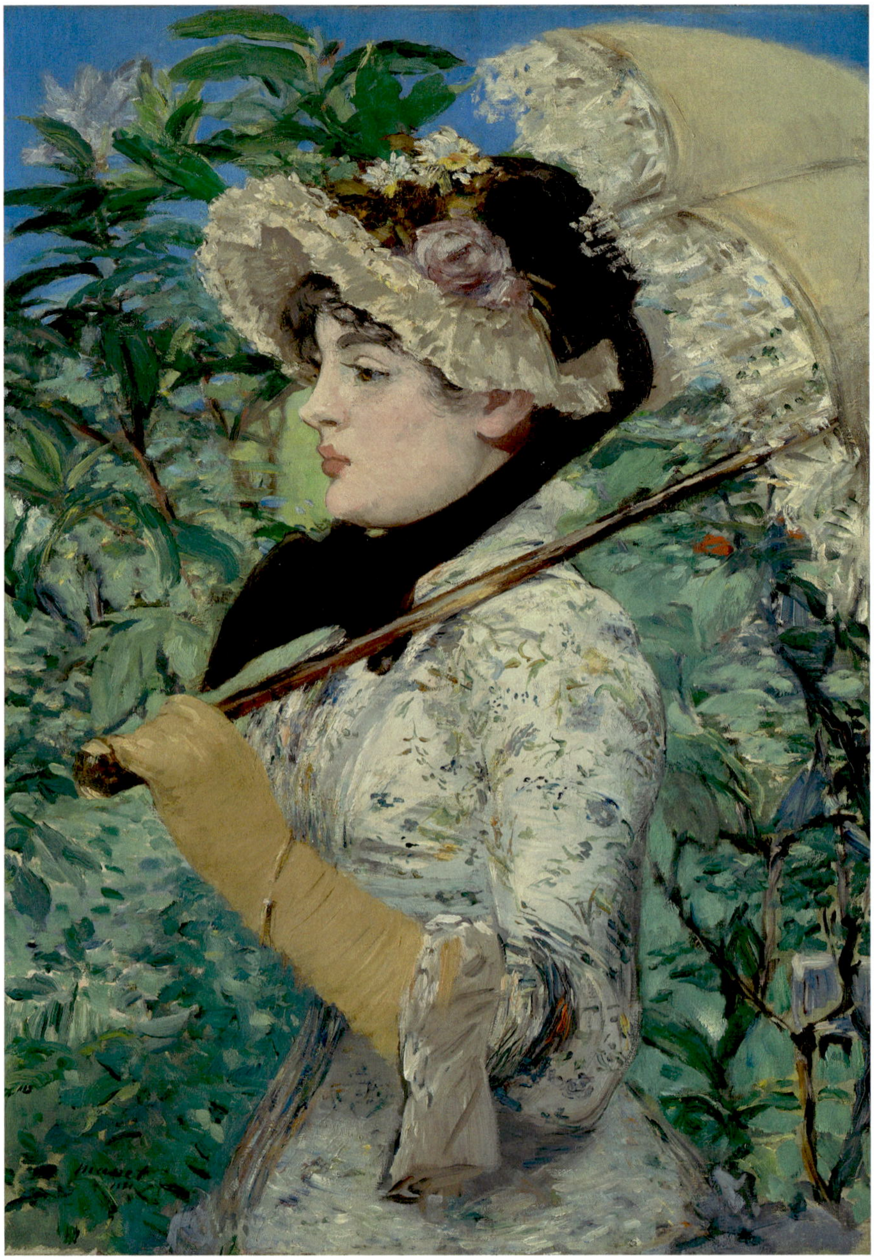

A

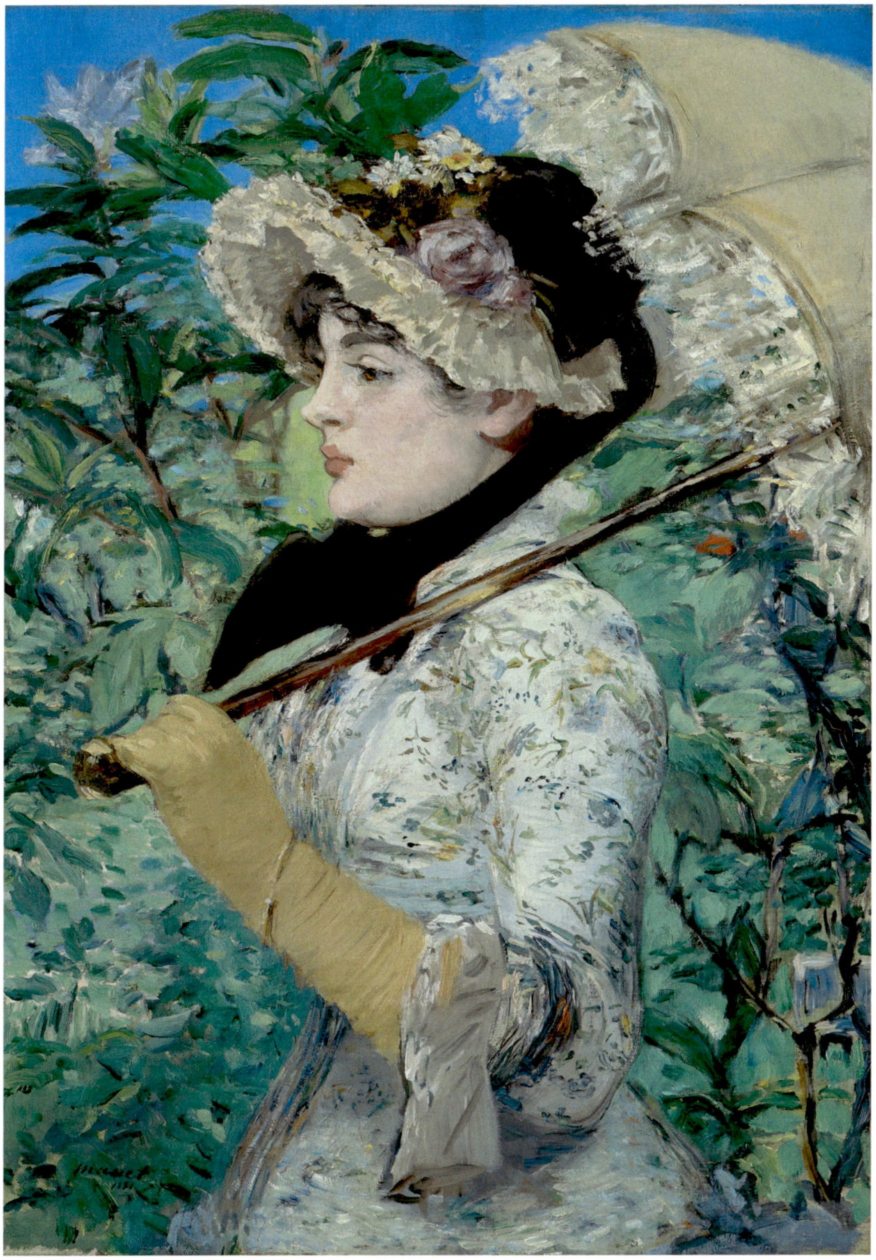

B

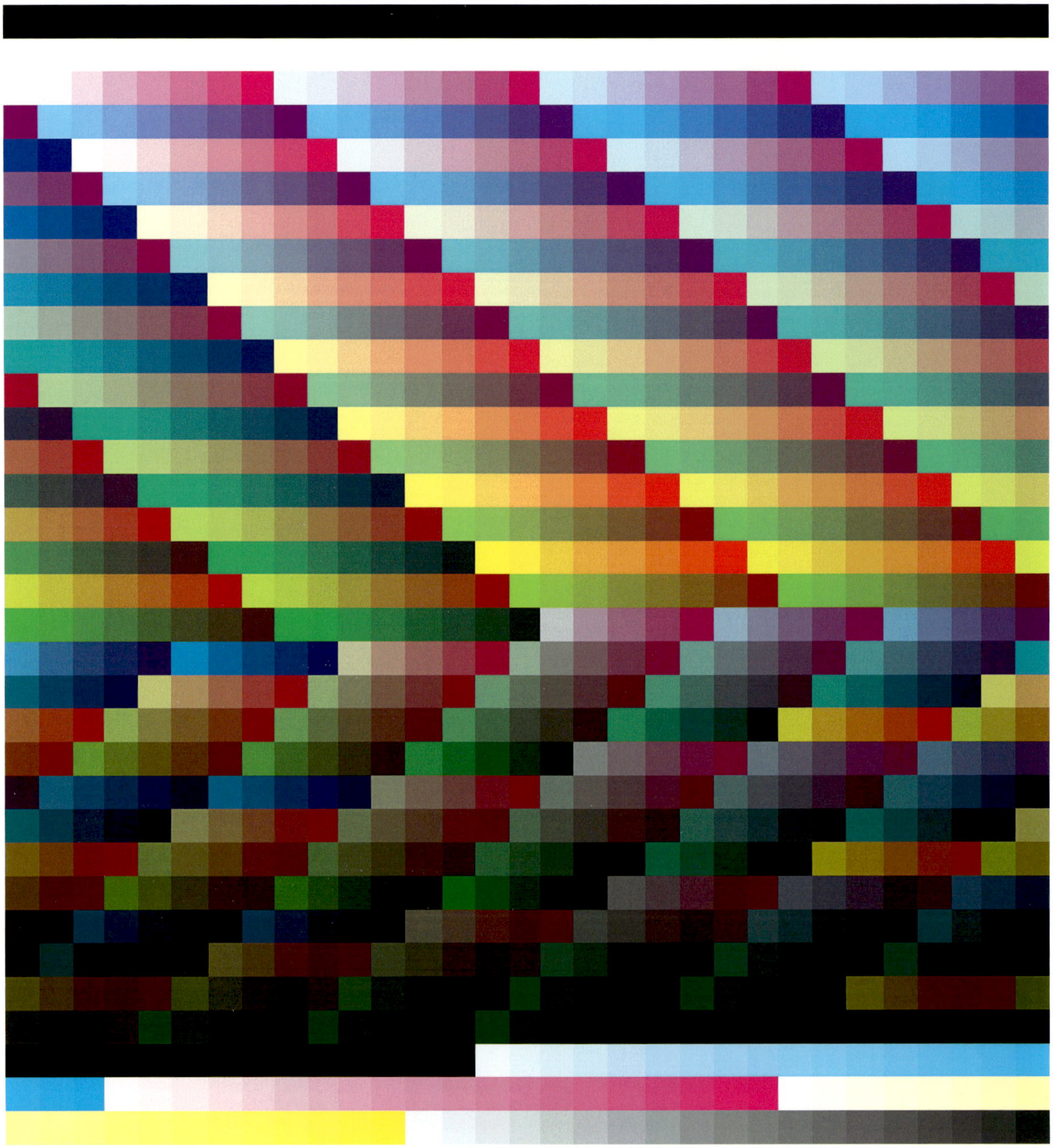

FIGURE 7.8. CMYK (cyan, magenta, yellow, and black) printer target used to create an ICC profile.

Why Reproducing Artwork Is Difficult

Achieving a high-quality reproduction or image archive is difficult. No matter who is doing the work and what their level of expertise, there are fundamental limitations preventing perfection, which in any case remains to be defined.

Consider an artwork imaging workflow consisting of a painting, its digitization, and its reproduction in print. The first decisions concern lighting. Should the print match the painting under gallery conditions, under the conditions experienced by the artist, in the photographic studio, or in the curator's office? Lighting dramatically affects the appearance of artwork because of color inconstancy. Suppose the painting is light sensitive and is exhibited under incandescent illumination at 2200 K below 50 lx. Under such conditions, the painting has a warm and slightly blurry appearance. Outside the exhibition is a point-of-sale kiosk lit by 5000 K natural daylight at an illuminance above 500 lx. If the gallery experience is reproduced for the kiosk lighting condition, the printed image will be yellowish, dark, and a bit blurry. Although the viewing experience is matched, the image has poor color and spatial image quality. A better approach might be to produce a print that matches the painting for the specific gallery lighting condition. That is, the print will match the painting when viewed adjacent to it. Certainly the exhibition curator will be pleased, knowing that the print matches the painting in the gallery. This is not common practice, however. Instead, following graphic arts practices, the painting and the print are viewed under standardized 5000 K daylight (CIE D50) at either 500 lx for practical appraisal or 2000 lx for critical appraisal (ISO 2009). Many museum imaging departments have such lighting conditions using high color-rendering daylight simulators to validate their color-management accuracy visually, and photographers will encourage curatorial and publications staff to approve any color decisions under these conditions. The photographic lighting should be similar in spectral properties to the defined reference illumination in order to reduce problems related to metamerism. Xenon flash lamps are recommended; other effective choices might be ceramic high-intensity discharge lamps, high-color-rendering solid-state lights with a 5000 K CCT (Berns 2014b), and incandescent lights with bluish filters.

The choice of camera is extremely important. Lens quality and sensor size play a larger role than the number of pixels in determining spatial image quality, as mentioned above.[9] The most important sensor property for color accuracy is its spectral sensitivity—the response of each color channel as a function of wavelength. Ideally, a color sensor should have spectral sensitivities that are equivalent to the eye's L, M, and S responses, plotted in figure 7.9.[10] The sensor will respond to incident light as our cone receptors do, and after processing, the images will be color accurate. This ideal is rarely met, however, because of manufacturing constraints and because color accuracy is only one of many design criteria. In particular, there is a tradeoff between color accuracy and image noise, and cameras designed for a wide range of lighting conditions may sacrifice color accuracy for spatial image quality (Kuniba and Berns 2009).

The inherent limitations in spectral sensitivity and the differences between the reference and camera-taking lighting are mitigated to some extent through color management. When a camera is "profiled," a reference target containing dozens or hundreds of color patches is imaged. Three such targets are shown in figure 7.10. Prior to imaging, the selected target is measured using a spectrophotometer, and reference CIELAB coordinates are calculated for the reference illuminant and observer. Software is used to create a camera ICC profile that minimizes color differences between the reference and camera CIELAB coordinates.[11] Profiles are most effective when the limitations in spectral sensitivity are not too severe, when the spectral properties of the target are similar to the work of art, and when the target's range of colors—its color gamut—exceeds that of the work of art.

The Digital ColorChecker SG target was used to create profiles for cameras having the spectral sensitivities plotted in figure 7.9. A xenon flash was used with the color-filter-array (CFA) camera, while tungsten halogen was used with the scanning camera. The Artist Paint Target was used to evaluate performance.[12] The results appear in figure 7.11. Both of the imaging systems had excellent lightness accuracy; the last row of neutrals matches the actual target well. The CFA system was much more accurate than the scanning system. Because the scanning system's red channel peaks at 630 nm and a tungsten halogen source was used, samples with appreciable reflectance at long wavelengths have color error, such as cobalt blue (row 2, column 2), pyrrole red (row 3, column 1), and quinacridone magenta (row 4, column 1). Cobalt blue becoming purplish has always been an acute problem with scanning cameras. The green samples were reduced in chroma. By comparison, the CFA system had much better color accuracy; only the reduction in chroma for the pyrrole red sample (row 3, column 1) is visible.

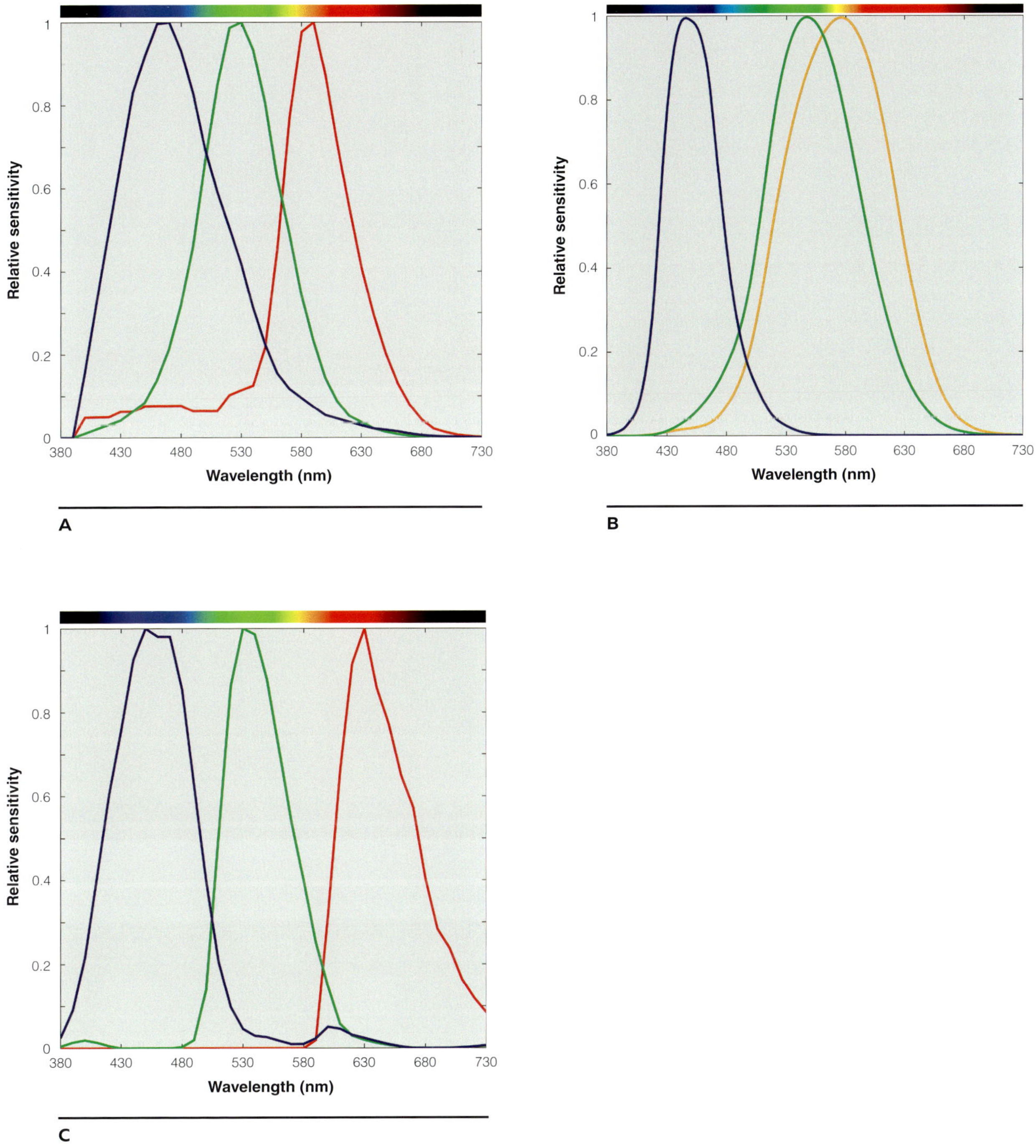

FIGURE 7.9. Relative spectral sensitivities of (A) a color-filter-array (CFA) camera system, (B) the human visual system, and (C) a scanning camera.

FIGURE 7.10. Typical targets used to create and evaluate camera profiles: (A) X-Rite Digital ColorChecker SG (20.5 × 28.8cm), (B) X-Rite ColorChecker Classic (20.5 × 28.8 cm), and (C) Artist Paint Target (6.4 × 8.6 cm).

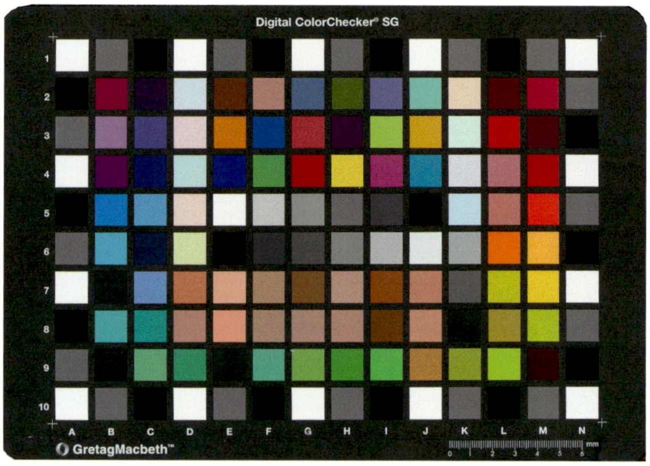

A

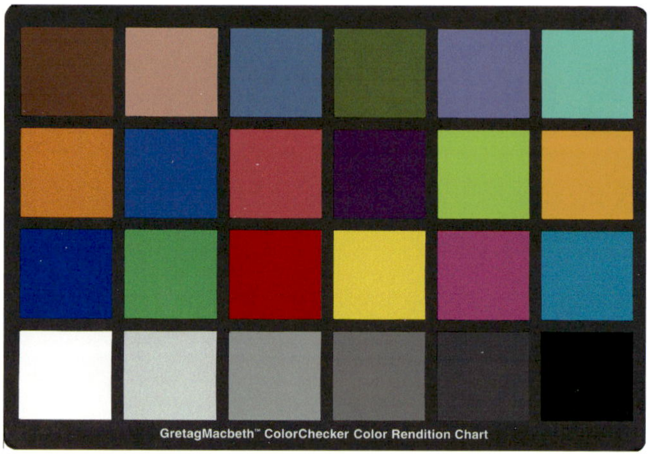

B

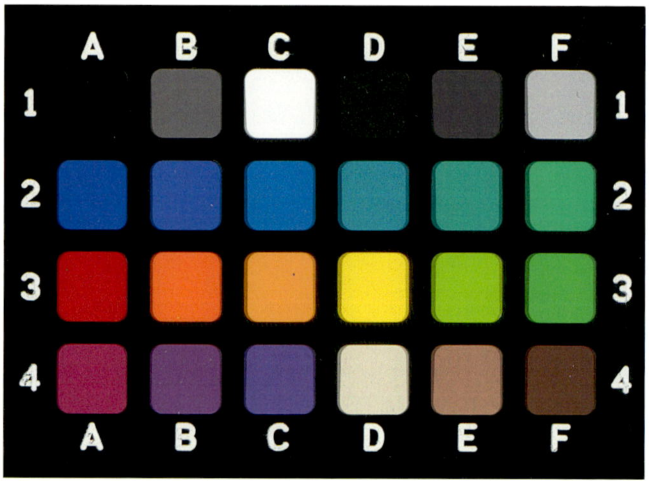

C

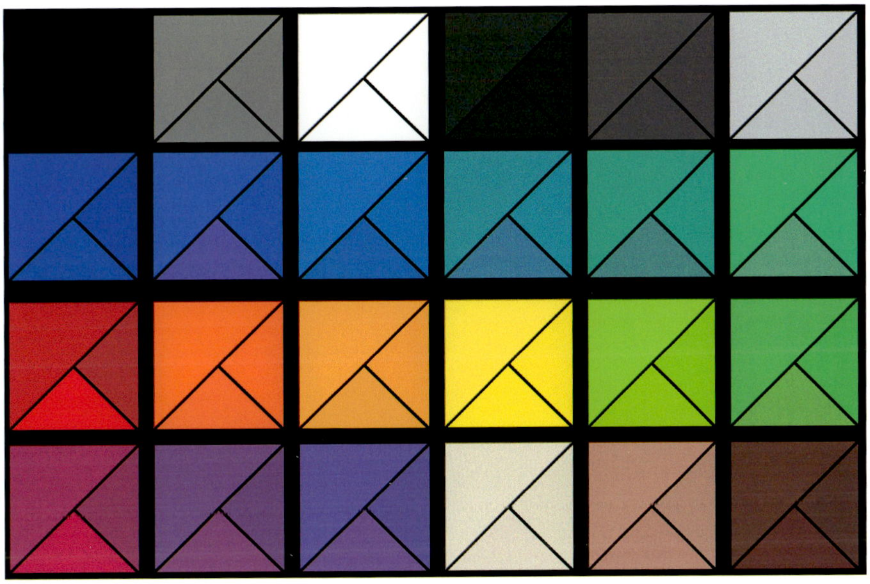

FIGURE 7.11. The Artist Paint Target (top left triangle) is compared with imaging using a color-managed CFA camera and xenon strobe lighting (right triangle) or a color-managed scanning camera and tungsten halogen lighting (bottom triangle).

Visual assessment, although very effective for qualitative evaluation, lacks precision for quantitative evaluation. In the latter case, it is more effective to calculate color differences and plot colors in CIELAB, as shown in figure 7.12. These "vector" plots point to the image colors of a color reproduction system. The color-coded dots indicate the location of the reference values, in this case based on direct spectrophotometry of the target. Ideally, the arrowheads and the colored dots are coincident. For both systems, the neutrals were imaged with high accuracy, which is the goal for color-managed camera systems used for artwork imaging (that is, colorimetric color reproduction). The CFA system had errors for three colors, mainly in chroma. The scanning system had large errors in both hue and chroma for most of the colors. Hue errors are indicated on the a*b* projection where the arrows point in directions other than toward or away from the origin (a* = b* = 0). Chroma errors are indicated by the length of the arrows in the $C^*_{ab}L^*$ plot. For the CFA system, the average and maximum color differences were 1.1 and 3.0 ΔE_{00} (2.8 and 10.7 ΔE^*_{ab}), respectively, while for the scanning system, they were 4.2 and 15.0 ΔE_{00} (9.0 and 26.0 ΔE^*_{ab}).[13] Only this CFA system has sufficient accuracy for imaging artwork (FADGI 2010; Dormolen 2012).

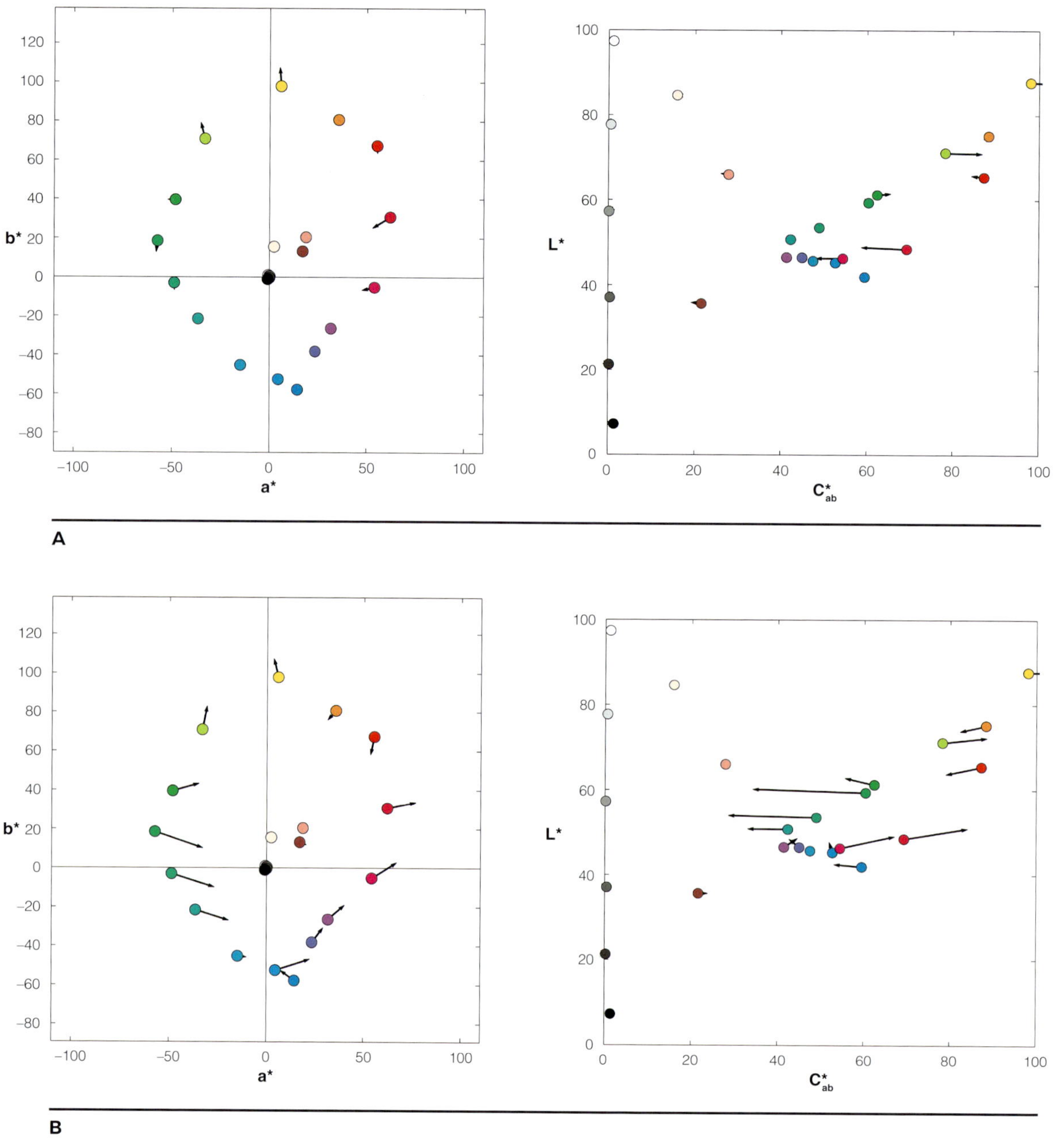

FIGURE 7.12. Color reproduction vector plots indicating the color accuracy of (A) CFA and (B) scanning camera systems when evaluating the Artist Paint Target. The colored dot defines the coordinates of the reference color, while the arrowhead defines the coordinates of the image-based colors.

Museum imaging professionals are quick to point out that reasonable performance for targets does not guarantee reasonable performance for paintings; as a consequence, image data are often manipulated in Photoshop in an effort to improve color accuracy through visual comparisons between the artwork and its rendering on a color-managed computer display. Unfortunately, this rarely does improve color accuracy (Berns et al. 2005; Frey and Farnand 2011). Because of mismatches in viewing conditions (the display is often dimmer than the illuminated painting), differences in image size, the need to look back and forth between the painting and display, and the tendency for preferred color reproduction, visually adjusted images often have increased chroma and lightness contrast compared with the painting (Newhall, Burnham, and Clark 1957; Hunt, Pitt, and Winter 1974; Nezamabadi, Berns, and Montag 2007). There will be instances in which a critical color is clearly wrong, necessitating visual correction. This should be done only for a well-profiled display with a white point and maximum luminance that match the painting's illumination.

The final step in this example is to produce a printed reproduction. There are two common approaches to profiling printing systems. One is to print a test target of known ink area coverages, measure each color patch with a spectrophotometer and record CIELAB values, and build a profile that relates CIELAB and area coverages. The profile is specific to the printer, paper, inks, color separation, and halftoning. This approach is most common with inkjet printers using ICC-based color management. The other approach is to calibrate a printing system to reproduce a standardized printing process[14] and to use standardized profiles (such as Uncoated FOGRA29 and U.S. Web Coated SWOP v2). In offset printing, this approach is used both for generating proofs and for the actual printing, facilitating the use of color management within a conventional printing environment.

A well-profiled or well-calibrated printer does not guarantee excellent results. There are inherent limitations with most printing systems. When the palette of the twenty-one artist paints is compared with offset printing that uses cyan, magenta, yellow, and black inks on smooth uncoated paper,[15] the differences in color gamut—shown in figure 7.13—are quite dramatic.[16] An offset-printed art book using these inks and paper cannot reproduce a modern artist's palette, particularly when a picture varnish is applied. On-demand inkjet printing is now used by museums to produce large reproductions, with the printer located in the museum

FIGURE 7.13. Sparsely sampled CIELAB coordinates of mixtures of twenty-one artist acrylic dispersion paints with a glossy picture varnish overlaid on (A) a CMYK offset print on smooth uncoated paper or (B) a CMYK inkjet print on high-quality inkjet paper. (The colors of the printer gamuts have been dulled.)

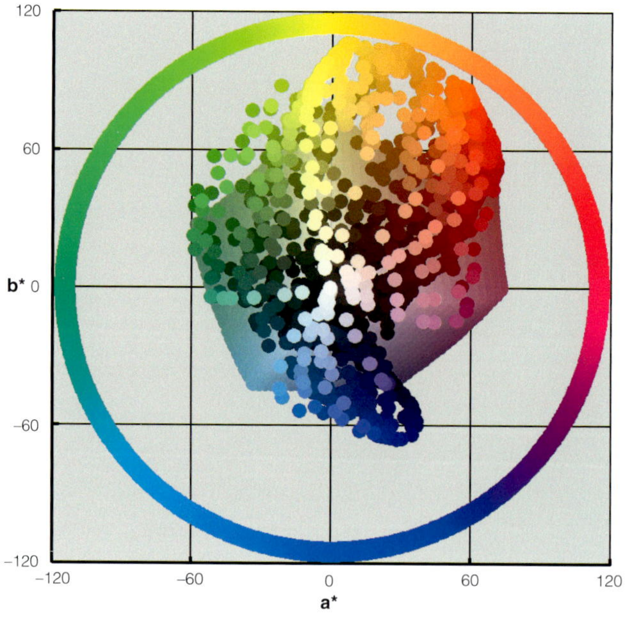

A

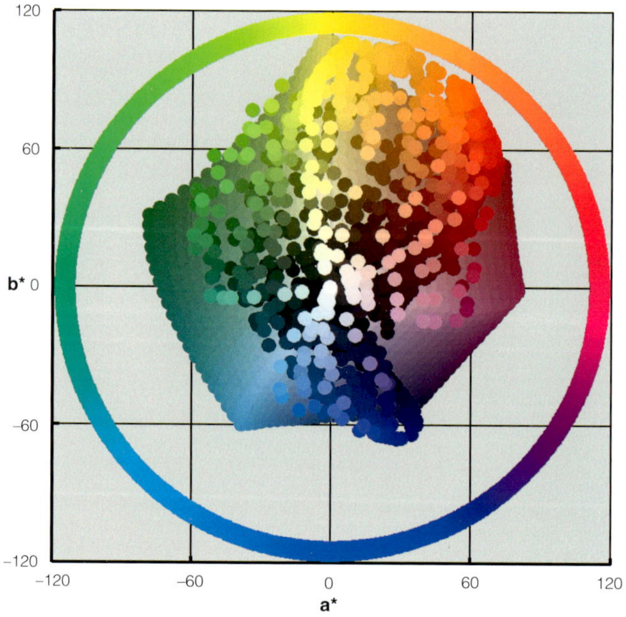

B

store. The color gamut for a four-color inkjet printer using high-quality paper with a small amount of optical brightener produces a gamut that encompasses many more of these colors, though not all of them. Printers with additional inks (such as red, green, blue, and orange) should produce color gamuts capable of reproducing most artwork.

The next limitation — metamerism — is the leading cause of acrimony between curators or artists and imaging professionals. The principles demonstrated in chapter 4 apply here as well. Pairs of matching colors produced from different colorants will mismatch if either the illumination or the observer changes. Since few, if any, artists paint with only cyan, magenta, yellow, black, and white paints, any print will be metameric to the artwork.

The difficulty is maintaining a color match when the artwork and print are viewed under a nonreference condition, as shown in figure 7.14. The metamer target introduced in chapter 4 was reproduced using a seven-ink inkjet printer.[17] Two prints were made, one using only cyan, magenta, and yellow and the other using all seven inks: cyan, magenta, yellow, black, red, green, and blue. For the three-ink print, ink amounts were calculated to result in a colorimetric match for CIE illuminant D50 and the CIE 1931 standard observer — the reference condition. For the seven-ink print, ink amounts were first calculated to give each sample the best spectral match, and then the amounts were adjusted slightly to produce a colorimetric match for the reference condition. By design, both prints matched all of the paint samples for the reference condition (fig. 7.14a).

Changing the illumination from D50 to 2200 K gallery lighting (fig. 7.14b) resulted in two kinds of mismatches, both examples of illuminant metamerism. First, the pairs of paint samples are metameric and, as a result, mismatch when illuminated by gallery lighting, as explained in chapter 4. Second, the three- and seven-ink prints are metameric compared with each painted sample. Both prints resulted in large errors. An interesting result was that only one sample of each pair was a poor match. In all cases, the paint sample contained pyrrole orange. If the multi-ink printer had contained an orange ink, the seven-ink prints would have matched both paint samples of each pair.

Changing the observer from the 1931 standard observer to the uncommon observer introduced in chapter 4 and illuminating with D50 also changes the reference condition. The resulting mismatches, similar in

FIGURE 7.14. The effect of changing the reference condition from (A) CIE illuminant D50 and the 1931 standard observer, to (B) 2000 K gallery lighting and the 1931 standard observer, or to (C) D50 and the uncommon observer. For each condition, the three-color print is shown in the top row, the paint samples in the middle, and the seven-color print at the bottom.

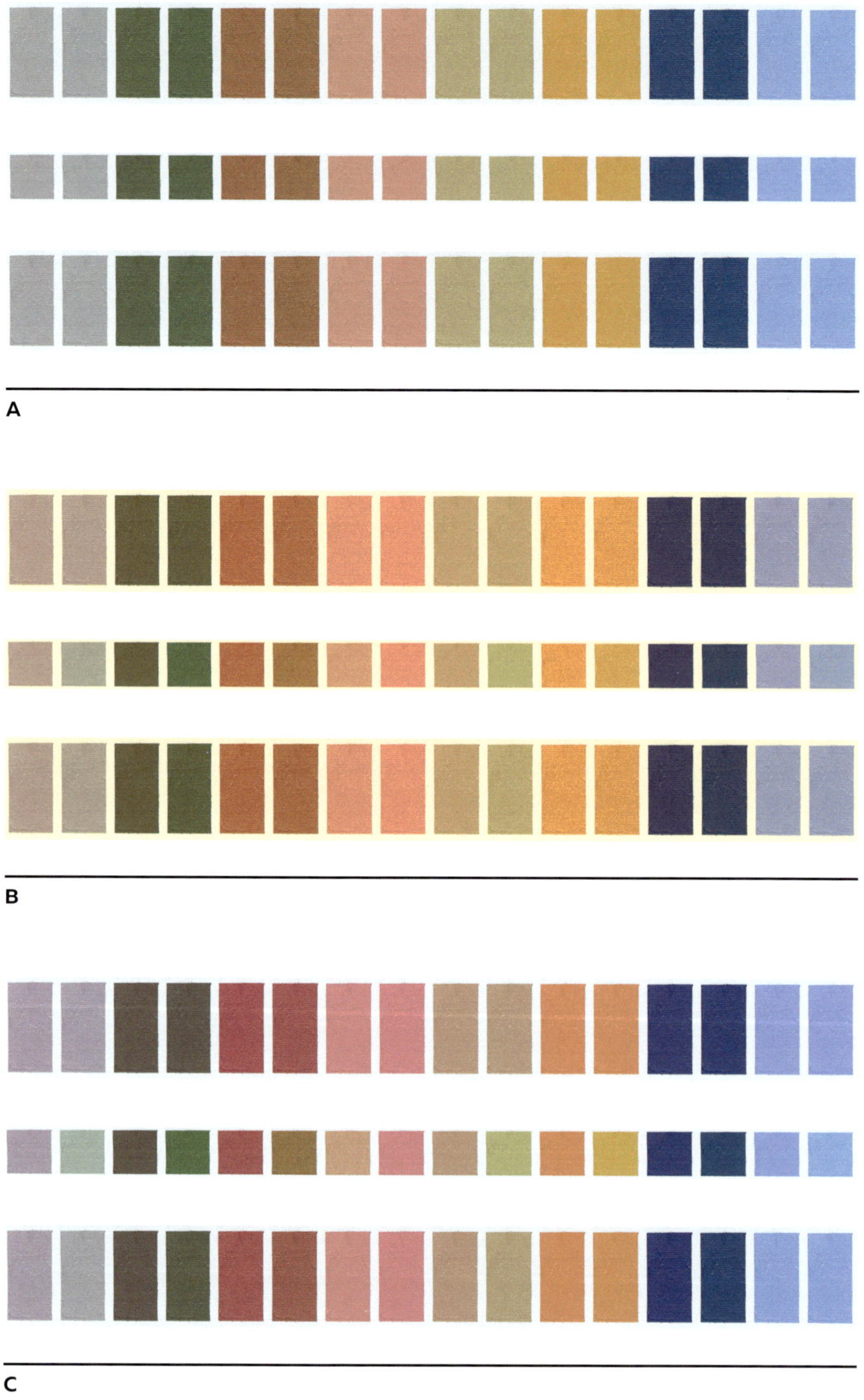

magnitude to those caused by changing the lighting, exemplify observer metamerism. Viewing the painting and proof in the imaging studio under reference lighting does not guarantee a match because a specific individual may have color vision that differs from that of the 1931 standard observer. As noted in chapter 4, the magnitude of color mismatch for any color-normal observer will vary between none—the standard observer—and that of the uncommon observer.

An additional problem, color inconstancy, is most apparent in the second paint sample in figure 7.14. Changing either the lighting or the observer altered its appearance from neutral gray to a low-chroma green. The sample was made using ultramarine blue, phthalocyanine green, pyrrole orange, and titanium white. As shown in figure 7.15, its reflectance spectrum is quite complex, leading to the significant color inconstancy (Berns, Billmeyer, and Sacher 1985). The three-color print's spectrum is also complex, though almost a mirror image of the paint sample; it, too, is color inconstant, but rather than becoming green, it turned to a low-chroma purple.

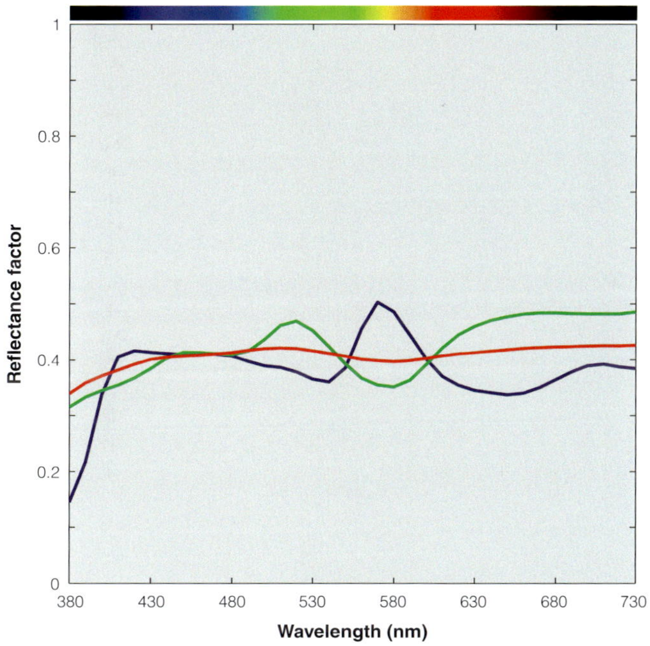

FIGURE 7.15. Reflectance spectra of the painted gray sample (blue line), three-ink print (green line), and seven-ink print (red line).

The seven-color print's spectrum is the least complex and, as a result, it is the least color inconstant, retaining a neutral appearance with changes in both lighting and observer. The best print might be the one that is least color inconstant rather than least metameric, particularly since printed reproductions such as catalogues, books, and posters are usually viewed without the original painting available for comparison (Chen et al. 2008).

Spectral Color Reproduction

Spectral color reproduction has many advantages compared with colorimetric color reproduction. A spectrally matching print is not metameric. A painting and print will match for all lighting conditions and all observers. The print can be used to evaluate how the painting will look under different lighting conditions, which is useful for scholarship and lighting design. A spectral imaging system would eliminate all the limitations of colorimetric color reproduction.

Despite these significant advantages, spectral reproduction is not available commercially. There are two reasons. First, ICC color management, which is universally used, is fundamentally three-dimensional with a colorimetric profile connection space. This is an insufficient amount of data to produce a nonmetameric print.[18] Second, there has not been a commercial demand for it.

The situation will soon change. Since 2014, the ICC has been developing an extension to PCS-based color management called iccMAX.[19] The PCS is replaced by a PCC, a profile connection condition, which facilitates spectral processing. This will enable multi- and hyperspectral cameras to connect directly with multi-ink printers and multi-primary displays. The advantages of spectral reproduction have already been demonstrated in the laboratory (Tsutsumi, Rosen, and Berns 2007, 2008; Berns, Taplin, et al. 2008; Urban, Rosen, and Berns 2008; Derhak and Berns 2010). An ICC spectral workflow will lead to dramatic improvements in color quality for imaging services. It will enable conservators and conservation scientists to analyze images from different imaging modalities without requiring computer-programming expertise, as iccMAX will include an image calculator. Images can also be defined as pigment maps, which will be useful for inpainting and studying an artist's working method.

Summary

Our ability to reproduce artwork on display and in print is a result of the trichromatic nature of vision, where hundreds of wavelengths are reduced to three color signals, defined using colorimetry. Assuming sufficient resolution, an original and a reproduction will match if the signals match for every pixel. Because display and printing systems use colorants that are rarely used by artists, these matches are metameric and it is necessary to standardize — in essence, restrict — the viewing conditions so that a calculated match actually matches for many observers. Departures from standardized conditions, such as viewing a printed reproduction in an incandescent-lit gallery adjacent to the artwork, often results in a severe mismatch.

Even when viewing is limited to standardized conditions, achieving color matches is still difficult because of limitations in camera design, printing inks, differences in color gamut, and observer metamerism. The proper use of color management mitigates these limitations to a great extent, though not completely. Spectral color reproduction holds great promise, eliminating many of the limitations of colorimetric color reproduction and providing new opportunities for the technical examination of art.

Notes

1. The image is being resampled using nearby pixels to approximate a higher-resolution sensor. As a mathematical approximation, it can never equal an optical equivalent.
2. This is the same reasoning as increasing resolution.
3. In graphic arts, the variable K is used for black ink. The origin of the K designation seems to be lost, but one possibility — the more likely one — is that K stood for "key." (In this interpretation, the other color separations were aligned with the "key" image, black, because it was the most visible.) Another possibility — the less likely one — is that K was chosen because it is the last letter of "black," and B was already being used to represent blue. (In the past, printers called their inks red, yellow, blue, and black.)
4. Information about the history of the ICC can be found at www.color.org.
5. When the color gamut is defined using CIELAB, L^* ranges from 0 to 100 and a^* and b^* ranges from −127 to 128. When the color gamut is defined using XYZ, each tristimulus value ranges from 0 to the illuminant's tristimulus value.
6. The term "color-rendering gamut" is used, rather than color gamut, because an encoding space is not a device that produces colors (Berns 2007).

7. Mixtures of the following Golden Artist Colors "Heavy Body" acrylic dispersion paints were produced computationally using optical data of each paint (Berns and Mohammadi 2007): cadmium yellow primrose (PY 35), bismuth vanadate yellow (PY 184), hansa (arylide) yellow opaque (PY 74), diarylide yellow (PY 83), cadmium orange (PO 20), pyrrole orange (PO 73), cadmium red light (PR 108), pyrrole red (PR 254), quinacridone red (PV 19), quinacridone magenta (PR 122), dioxazine purple (PV 23), ultramarine blue (PB 29), cobalt blue (PB 28), cerulean blue, chromium (PB 36:1), phthalo blue (red shade) (PB 15:1), phthalo blue (green shade) (PB 15:4), phthalo green (blue shade) (PG 7), phthalo green (yellow shade) (PG 36), ivory black (PBk 7), and titanium white (PW 6). A high-gloss varnish was applied, also computationally (Abed, Berns, and Masaoka 2013).
8. Photoshop refers to these as perceptual, saturation, relative colorimetric, and absolute colorimetric.
9. For book reproduction, 300 pixels per inch (ppi) is generally sufficient for pictorial images (Hunt 2004). For line art and text, 600 ppi is a better choice. For this book, both pictorial images and graphs were 600 ppi and encoded as ProPhoto RGB 16-bit TIFF files. The increased image resolution led to more flexibility in layout design.
10. The term "equivalent" was used deliberately. Equivalent spectral sensitivities are sensitivities that, following a linear transformation, are identical to LMS responses, known as the Ives or Luther-Ives criterion (Berns 2001).
11. Each company that offers software to build color profiles uses its own approach and in some cases its own targets. As a consequence, color accuracy varies, even for identical cameras.
12. This is known as independent data, that is, data not used to build the profile. Quite often, color accuracy is reported based on the profiling target. It is important to evaluate independent data.
13. Both camera systems were computational: the spectral data for the ColorChecker SG and Artist Paint Target were used to calculate camera signals given each device's spectral sensitivities and taking-illuminant spectral radiance. Because there was no image noise, the color error is smaller than in real imaging experiments (Berns and Smith 2012).
14. Standards exist for many different printing technologies, defined by the International Organization for Standardization (ISO) Technical Committee on Graphic Technology, TC 130.
15. The color gamut of smooth uncoated paper printed with an offset press was calculated using spectral measurements of each ink and the paper of *Billmeyer and Saltzman's Principles of Color Technology* (Berns 2000).
16. When a significant amount of color gamut mapping is required, the customer (e.g., artist, curator, etc.) may want to perform the mapping rather than use the embedded mapping in the printer profile or relying on the printer when performing color separation. Using

Photoshop's proofing feature and the print profile, the image is adjusted until colors are within the print color gamut. However, soft proofing has limited accuracy, and for very color-critical requirements, printed proofs should be produced.

17. The printer was an HP Designjet Z3100 with HP semigloss proofing paper.
18. There is one exception. If the painting was produced using only three chromatic paints plus white, and a three-color print was produced using inks with the identical spectral properties to the paints, a colorimetric match would also be a spectral match.
19. More information can be found at www.color.org.

References

Abed, F. M., R. S. Berns, and K. Masaoka. 2013. "Geometry-Independent Target-Based Camera Colorimetric Characterization." *Journal of Imaging Science and Technology* 57: 50503-1–50503-15.

Berns, R. S. 2000. *Billmeyer and Saltzman's Principles of Color Technology*. 3rd ed. New York: Wiley Interscience.

———. 2001. "The Science of Digitizing Paintings for Color-Accurate Image Archives: A Review." *Journal of Imaging Science and Technology* 45: 373–83.

———. 2007. "Let's Call It 'Color-Gamut Rendering.'" *Color Research and Application 32:* 334–35.

———. 2014a. "Camera Encoding Evaluation for Image Archiving of Cultural Heritage." Technical report, Studio for Scientific Imaging and Archiving of Cultural Heritage, Rochester Institute of Technology. http://www.rit-mcsl.org/Mellon/PDFs/CameraEncoding_TR_May_2014.pdf.

———. 2014b. "Evaluating Solid State and Tungsten-Halogen Lighting for Imaging Artwork via Computer Simulation." Technical report, Studio for Scientific Imaging and Archiving of Cultural Heritage, Rochester Institute of Technology. http://www.rit-mcsl.org/Mellon/PDFs/Evaluating_Solid_State_Lighting_TR_Jan_2014.pdf.

Berns, R. S., J. F. W. Billmeyer, and R. S. Sacher. 1985. "Methods for Generating Spectral Reflectance Functions Leading to Color-Constant Properties." *Color Research and Application* 10: 73–83.

Berns, R. S., and M. Derhak. 2014. "A New Encoding System for Image Archiving of Cultural Heritage: ETRGB." In *Archiving 2015,* 74–77. Los Angeles: Society for Imaging Science and Technology.

Berns, R. S., F. S. Frey, M. R. Rosen, E. P. Smoyer, and L. A. Taplin. 2005. *Direct Digital Capture of Cultural Heritage: Benchmarking American Museum Practices and Defining Future Needs Final Report*. Rochester: Rochester Institute of Technology. http://www.art-si.org/PDFs/Metric/Benchmark_Report_April.pdf.

Berns, R. S., and M. Mohammadi. 2007. "Evaluating Single- and Two-Constant Kubelka-Munk Turbid Media Theory for Instrumental-Based Inpainting." *Studies in Conservation* 52: 299–314.

Berns, R. S., and S. Smith. 2012. "Analysis of Color Management Default Camera Profiles for Museum Imaging Applications." In *Archiving 2012,* 111–15. Copenhagen: Society for Imaging Science and Technology.

Berns, R. S., L. A. Taplin, P. Urban, and Y. Zhao. 2008. "Spectral Color Reproduction of Paintings." In *CGIV 2008—Fourth European Conference on Colour in Graphics, Imaging, and MCS/08 Vision 10th International Symposium on Multispectral Colour Science,* 484–88. Barcelona: Society for Imaging Science and Technology.

Chen, Y., R. S. Berns, L. A. Taplin, and F. H. Imai. 2008. "A Multi-Ink Color-Separation Algorithm Improving Image Quality." *Journal of Imaging Science and Technology* 52: 20604-1–20604-9.

Derhak, M. W., and R. S. Berns. 2010. "Comparing LABPQR and the Spectral Gamut Mapping Framework." In *Eighteenth Color and Imaging Conference: Color Science and Engineering Systems, Technologies, and Applications,* 206–12. San Antonio: Society for Imaging Science and Technology.

Dormolen, H. van. 2012. *Metamorfoze Preservation Imaging Guidelines: Image Quality, Version 1.0, January 2012*. The Hague: National Library of the Netherlands.

Evans, R. M. 1959. *Eye, Film, and Camera in Color Photography*. New York: Wiley.

FADGI (Federal Agencies Digitization Initiative, Still Image Working Group). 2010. *Technical Guidelines for Digitizing Cultural Heritage Materials: Creation of Raster Image Master Files*. Washington, DC: U.S. National Archives and Records Administration.

Frey, F. S., and S. Farnand. 2011. *Benchmarking Art Image Interchange Cycles*. Rochester: Rochester Institute of Technology, http://www.rit-mcsl.org/Mellon/PDFs/benchmarking_art_image_interchange_cycles.pdf.

Green, P. 2010. *Color Management: Understanding and Using ICC Profiles*. Chichester, UK: Wiley.

Hunt, R. W. G. 1970. "Objectives in Colour Reproduction." *Journal of Photographic Science* 18: 205–15.

———. 2004. *The Reproduction of Colour*. 6th ed. Chichester, UK: Wiley.

Hunt, R. W. G., I. T. Pitt, and L. M. Winter. 1974. "The Preferred Reproduction of Blue Sky, Green Grass and Caucasian Skin in Colour Photography." *Journal of Photographic Science 22: 144–50.*

IEC. *1999. Multimedia Systems and Equipment: Colour Measurement and Management. Part 2-1: Colour Management—Default RGB Colour Space—sRGB (61966-2-1 ed1.0)*. Geneva: International Electrotechnical Commission.

ISO. 2009. *Graphic Technology and Photography: Viewing Condition (3664:2009)*. Geneva: International Organization for Standardization.

Kuniba, H., and R. S. Berns. 2009. "Spectral Sensitivity Optimization of Color Image Sensors Considering Photon Shot Noise." *Journal of Electronic Imaging* 18: 023002-1–023002-14.

Morovic, J. 2008. *Color Gamut Mapping, Wiley-IS&T Series in Imaging Science and Technology*. New York: Wiley.

Newhall, S. M., R. W. Burnham, and J. R. Clark. 1957. "Successive and Simultaneous Color Matching." *Journal of the Optical Society of America* 47: 43–56.

Nezamabadi, M., R. S. Berns, and E. D. Montag. 2007. "An Investigation of the Effect of Image Size on the Color Appearance of Softcopy Reproductions Using a Contrast Matching Technique." In *Color Imaging XII: Processing, Hardcopy, and Applications,* 649309. San Jose: SPIE.

Tsutsumi, S., M. R. Rosen, and R. S. Berns. 2007. "Spectral Gamut Mapping Using LabPQR." *Journal of Imaging Science and Technology* 51: 473–85.

——. 2008. "Spectral Color Management Using Interim Connections Spaces Based on Spectral Decomposition." *Color Research and Application* 33: 282–99.

Urban, P., M. Rosen, and R. S. Berns. 2008. "Spectral Gamut Mapping Framework Based on Human Color Vision." In *CGIV 2008 – Fourth European Conference on Colour in Graphics, Imaging, and MCS/08 Vision 10th International Symposium on Multispectral Colour Science,* 548–53. Barcelona: Society for Imaging Science and Technology.

Yendrikhovskij, S. N., F. J. J. Blommaert, and H. de Ridder. 1999. "Color Reproduction and the Naturalness Constraint." *Color Research and Application* 24: 52–67.

FURTHER READING

Many books were used as background reading and as reference material in the preparation of the present volume. I found a number of them particularly insightful and recommend them for further study.

Arnkil, H. 2013. *Colours in the Visual World*. Helsinki: Aalto ARTS Books.

Color is multidisciplinary. Over the years, books have appeared that consider color from multiple perspectives. Harald Arnkil's contribution to the literature is reasonably successful. Originally published in Finnish in 2007, this book "is intended for studying colour and visual phenomena in art, design, and architecture," as the foreword explains. The layout is visually stunning, perfect for the content. Twelve chapters, each with a set of assignments, encompass physiology, physics, and psychology. The book would fit nicely into a non-majors course, fulfilling a liberal arts requirement. For the specialist, however, there are too many gaps, and some of the information is incorrect or obsolete.

Berns, R. S. 2000. *Billmeyer and Saltzman's Principles of Color Technology*. 3rd ed. New York: Wiley Interscience.

This book is used as an introductory text in academia, industry, and art conservation science. Fred Billmeyer Jr. and Max Saltzman, highly respected and accomplished chemists as well as color scientists, wrote the first two editions with the goal of producing an authoritative text that was accessible to color technologists without extensive mathematical preparation. Its seven chapters deal with defining color, describing color, measuring color, measuring color quality, colorants, producing colors, and applying the book's content in practice. In the third edition, an appendix was added on the mathematics of color technology, enabling its use in a color science college curriculum.

Boyce, P. R. 2003. *Human Factors in Lighting*. 2nd ed. New York: Taylor & Francis.

This is a comprehensive and authoritative book about lighting from a visual perspective. It is divided into three parts: fundamentals (light, the visual system, and the circadian system), generalities (lighting and work, lighting and visual discomfort, and lighting and the perception of spaces and objects), and specifics (such as lighting for offices, escape lighting, and lighting and crime).

Cuttle, C. 2007. *Light for Art's Sake: Lighting for Artworks and Museum Displays*. Amsterdam: Butterworth-Heinemann.

This text from Christopher Cuttle, a world expert on museum lighting, is a tour de force, a culmination of the author's extensive experience and well-reasoned opinions on museum lighting. Cuttle uses in situ images from museums and galleries around the world to illustrate concepts. The background material on physics, perception, and lighting geometry is very accessible. This is an excellent resource for understanding the tradeoffs between the appearance of the art, a gallery's aesthetics, and conservation requirements.

Daw, N. 2012. *How Vision Works: The Physiological Mechanisms behind What We See*. New York: Oxford University Press.

This is an excellent introductory text about human vision with a focus on its physiology. There are twelve chapters: introduction, organization of the visual system, brightness and contrast, color vision, perception of motion, depth perception, objects and faces, control of eye movements, adaptation and aftereffects, attention, visual memory, and summary. The balance of text, scientific figures, and visual examples keeps the reading interest high. The references are extensive and would provide a useful point of entry to the primary literature.

Dilaura, D. L., K. W. Houser, R. G. Mistrick, and G. R. Steffy, eds. 2011. *The Lighting Handbook: Reference and Application*. 10th ed. New York: Illuminating Engineering Society of North America.

This handbook is highly valuable to both illuminating engineers and lighting designers. Its thirty-seven chapters are divided into three sections: framework, design, and applications. The preface explains: "Framework chapters describe the science and technology related to lighting, including vision, optics, nonvisual effects of optical radiation, photometry, and light sources. Design chapters include not only fundamental considerations and special issues of daylighting and electric lighting design, but also energy management, controls, and economics. Applications chapters establish the design context for many lighting applications, provide illuminance recommendations for specific tasks and areas, and identify some of the analytic goals of lighting design using science and technology." There is even a chapter on lighting for art.

Elliot, A. J., M. D. Fairchild, and A. Franklin, eds. 2015. *Handbook of Color Psychology*. Cambridge: Cambridge University Press.

This contributed chapter book represents the current understanding of color, from the physical world, through our physiology, to our mind's eye. There are thirty-four chapters divided into nine parts: introduction; basics of color science; development of and differences in color vision; color categorization; color symbolism and association; color preferences; color effects on psychological and biological functions; psychological effects on color perception; and color phenomena. With a remarkable breadth in the coverage of topics, this book is an important addition to the library of any scholar in a color-related discipline.

Fairchild, M. D. 2013. *Color Appearance Models*. 3rd ed. Chichester, UK: Wiley.

Color appearance models extend traditional trichromatic and opponent color vision by including adaptation, surround, background, and similar effects on perceived color. This topic is the author's main research area, and this book is very readable and authoritative, explaining many visual phenomena—in effect, bridging the gap between Newton and Goethe.

Fraser, B., C. Murphy, and F. Bunting. 2005. *Real World Color Management*. 2nd ed. Berkeley: Peachpit.

In the world of color management, 2005 is long ago and far away. Nonetheless, this remains the best book about color management. It is divided into four parts: introduction to color management, building and tuning profiles, applications and workflow, and appendices. The chapter on color management is a triumph of clarity.

Hunt, R. W. G. 2004. *The Reproduction of Colour*. 6th ed. Chichester, UK: Wiley.

This comprehensive and authoritative book about the theory and practice of color reproduction is divided into six parts: fundamentals, color photography, color television, color printing, digital imaging, and evaluating color appearance. A comparison of all editions, beginning with the first, in 1957, reveals the evolution of commercial color-imaging systems. Robert Hunt has been a leader in color imaging during most of his sixty-plus-year career. This book is a true classic.

Johnston-Feller, R. 2001. *Color Science in the Examination of Museum Objects: Nondestructive Procedures*. Tools for Conservation. Los Angeles: Getty Conservation Institute.

Ruth Johnston-Feller had a long and distinguished career as a dye chemist and color scientist with expertise in instrumental-based color matching. From the 1970s on, she applied her expertise to problems of art conservation, culminating in this book. For readers with a color science background, this book is a treasure trove of information. Her insights into the color science of museum objects are unparalleled.

Kirsh, A., and R. S. Levenson. 2000. *Through Paintings: Physical Examination in Art Historical Studies*. Materials and Meaning in the Fine Arts, vol. 1. New Haven, CT: Yale University Press.

This introductory book on the technical examination of paintings is written very clearly, and the many color illustrations are excellent. The book is divided into six chapters: introduction, the support, the ground and preparatory layers, the paint layer, the varnish layer, and beyond the painting.

Kuehni, R. G. 2013. *Color: An Introduction to Practice and Principles*. 3rd ed. New York: Wiley.

It is rare for an author to have expertise in color science, colorants, color mixing, color-order systems, and color history, but Rolf Kuehni does, and as a result has written a book that is authoritative, comprehensive, and insightful, and one that is far more readable than books with chapters contributed by many people. Topics include sources of color, what color is and how we have come to experience it, from light to color, color perception phenomena, orderly arrangements of color, defining the color stimulus, calculating color, colorants and their mixture, color reproduction, the web of color, color (theory) in art, and harmony of colors. There is also an appendix with a timetable of color in science and art as well as a glossary.

Kuehni, R. G., and A. Schwarz. 2008. *Color Ordered: A Survey of Color Order Systems from Antiquity to the Present*. Oxford: Oxford University Press.

This beautiful book is an encyclopedic historical survey of systematic arrangements of color. The book is well organized and enables the reader to understand how color theory and physical embodiments of these theories have evolved since antiquity. There are full-color plates of many of the color systems at the end of the book plus a chronological timeline of each color system cross-referenced by chapter and page number, and a glossary. This book is for anyone with an interest in the history of color.

Lamb, T., and J. Bourriau, eds. 1995. *Colour: Art and Science*. Darwin College Lectures. Cambridge, UK: Cambridge University Press.

This is a contributed chapter book written by well-known experts in color vision, color physics, art conservation, culture, language, and easel painting. It is the classic comprehensive coverage of color, art, and science.

Livingstone, M. 2014. *Vision and Art: The Biology of Seeing*. Updated and expanded edition. New York: Harry N. Abrams.

The biology of vision is explained using paintings and drawings as illustrative examples. This update includes several additional chapters and a significant improvement in its organization. There are seventeen chapters explaining color vision, spatial vision, producing 3-D on a 2-D surface such as a painting, color physics, and talent and learning disabilities. The book is very accessible.

MacDonald, L., ed. 2006. *Digital Heritage: Applying Digital Imaging to Cultural Heritage*. Amsterdam: Elsevier.

This is a contributed chapter book focusing on imaging cultural heritage. Its editor, Lindsay MacDonald, has considerable expertise in both color reproduction and cultural heritage imaging. The twenty chapters are divided into three parts: user requirements, technology, and applications. The first and third parts are especially interesting, as these subject areas are rarely included along with technical aspects of digital imaging.

Nassau, K., ed. 1998. *Color for Science, Art and Technology*. Azimuth, vol. 1. Amsterdam: Elsevier.

"The aim of this book is to assemble a series of chapters, written by experts in their fields, covering the basics of color—and then some," the editor, Kurt Nassau, explains. The contributed chapters are organized in three sections: the science of color; color in art, culture, and life; and colorants and the preservation and reproduction of color. Each chapter is a topical overview with ample text and references. The amount of mathematics is modest, as is the number of color plates. The best aspect of the book is the authors' insights into their areas of expertise.

Nassau, K. 2001. *The Physics and Chemistry of Color: The Fifteen Causes of Color*. 2nd ed. Wiley Series in Pure and Applied Optics. New York: Wiley Interscience.

This is an excellent book for gaining a thorough understanding of how color is produced. It is divided into seven parts: light and color, color involving vibrations and simple excitations, color involving ligand field effects, color involving molecular orbitals, color involving band theory, color involving geometrical and physical optics, and color-related topics. There are color figures of the fifteen causes of color. The book assumes a secondary-school understanding of physics and chemistry.

Rood, O. N. 1879. *Modern Chromatics, with Applications to Art and Industry*. International Scientific Series. New York: D. Appleton.

Written by the American physicist Ogden Rood and translated into French and German, this is the most influential color science text of the nineteenth century. The French edition had a profound impact on the Neo-Impressionists. Steeped in Young-Helmholtz trichromatic theory and a keen observer of well-controlled color-mixing experiments using spinning disks, Rood explains color physics, color perception, trichromatic theory, color mixing, and color-order systems. He even includes a page on the new theory of color vision by Ewald Hering.

Rossing, T. D., and C. J. Chiaverina. 1999. *Light Science: Physics and the Visual Arts*. Undergraduate Texts in Contemporary Physics. New York: Springer.

This is an introductory physics book, focusing on topics of interest to the visual arts. Each chapter contains a summary, references, a glossary, review questions, and suggested experiments. It is written clearly and concisely in a nonmathematical fashion.

Rossotti, H. 1985. *Colour: Why the World Isn't Grey*. Princeton Science Library. Princeton, NJ: Princeton University Press.

This book is excellent and surprising, providing the reader with direct and concise information about color physics, color perception, color technology, and color psychology. It is nonmathematical; drawings are used to convey color physics and physiology. The information is accurate relative to its publication date of 1985. One wonderful feature of the book is its collection of quotations from artists, philosophers, physicists, and poets. The surprise is that Hazel Rossotti is able to explain why the world isn't gray without using color: the book is a black-and-white paperback.

Stoner, J. H., and R. Rushfield, eds. 2012. *The Conservation of Easel Paintings*. Routledge Series in Conservation and Museology. New York: Routledge.

Every conservator and conservation scientist should have access to this outstanding book, which covers other media in addition to paintings. More than seventy-five authors contributed to forty-nine chapters, divided into three parts. The first deals with technical art history, examination, documentation, and scientific analysis; the second, methods and approaches for the treatment and care of easel paintings; and the third, preventive conservation, health and safety, outreach, and professional organizations.

Taft, W. S., J. W. Mayer, R. Newman, D. Stulik, and P. I. Kuniholm. 2000. *The Science of Paintings*. New York: Springer.

This is an introductory book about the science of paintings, often used in art conservation curricula. Chapters include the structure and analysis of paintings, paint, organic binders, the painter's color and light, color and light, optics of paint films, beyond the eye, and detection of fakes, and there are also a number of appendices with details for scientific analysis of paintings. The physical and chemical aspects of paintings are well described. The perceptual aspects, however, are not well presented and include common errors such as confusing metamerism and color inconstancy.

Thomson, G. 1986. *The Museum Environment*. 2nd ed. Butterworth-Heinemann Series in Conservation and Museology. Oxford: Butterworth-Heinemann.

Written by the scientific advisor of twenty-five years at the National Gallery, London, this is the seminal text on issues relating to lighting, humidity, and air pollution. The book has two parts. The first is qualitative, revealing Garry Thomson's considerable expertise and experience. The writing style is engaging and easy to follow. The second part is quantitative, giving methods for calculations and more technical detail.

Williamson, S. J., and H. Z. Cummins. 1983. *Light and Color in Nature and Art*. New York: John Wiley and Sons.

This is a classic text for a non-majors course in physics, written by two physics professors who are well versed in vision and in artist colorants. The book is well written, concise, and filled with useful content. It has two drawbacks: first, it limits colorimetry to tristimulus values and chromaticities; and second, because of its publication date, it does not consider digital imaging.

Zollinger, H. 1999. *Color: A Multidisciplinary Approach*. New York: Wiley-VCH.

Seven excellent chapters lead the reader from color physics through color and art, concluding with an epilogue that ties the topics together and provides interesting insights from the author, a dye chemistry expert with a passion for art and culture. The book does not attempt to be comprehensive; rather, it is multidisciplinary, as stated in the subtitle. Chapters include an introduction, physics of light and color, chemistry of color, colorimetry, how we see colors, how we name colors, color in art and other cultural activities, and an epilogue. The book is mainly textual, with color illustrations of artwork.

Zwimpfer, M. 1988. *Color, Light, Sight, Sense: An Elementary Theory of Color in Pictures*. West Chester, PA: Schiffer.

This is a remarkable introductory book about color. It is divided into chapters on light, color, ways in which colors appear, the eye, colors created in the eye, color designation and organization, effect and significance of colors, and technical reproduction of colors. The annotated illustrations—more than five hundred of them—are the teaching tool. The author was a design teacher; hence his clear, uncluttered illustrations. He has used his obviously deep understanding of color printing to produce excellent examples of color mixing.

INDEX

Note: Page numbers in italic type indicate figures or illustrations.